The aim of the Handbooks in Practical Animal Cell Biology is to provide practical workbooks for those involved in primary cell culture. Each volume addresses a different cell lineage, and contains an introductory section followed by individual chapters on the culture of specific differentiated cell types. The authors of each chapter are leading researchers in their fields and use their first-hand experience to present reliable techniques in a clear and thorough manner.

Epithelial Cell Culture contains chapters on epithelial cells derived from airway, intestine, pancreas, kidney and bladder, genital ducts, mammary glands, keratinocytes, and skin glands and appendages.

Epithelial cell culture

Epithelial cell culture

Edited by

Ann Harris

Lecturer in Paediatric Molecular Genetics,
Institute of Molecular Medicine,
University of Oxford, Oxford, UK

Published by the Press Syndicate of the University of Cambridge
The Pitt Building, Trumpington Street, Cambridge CB2 1RP
40 West 20th Street, New York, NY 10011-4211, USA
10 Stamford Road, Oakleigh, Melbourne 3166, Australia

First published 1996

Printed in Great Britain at the University Press, Cambridge

A catalogue record for this book is available from the British Library

Library of Congress cataloguing in publication data

Epithelial cell culture / edited by Ann Harris.
p. cm.
Includes index.
ISBN 0 521 55023 8 (hardcover). – ISBN 0 521 55991 X (pbk.)
1. Epithelial cells – Laboratory manuals. 2. Epithelium – Cultures and culture media – Laboratory manuals. I. Harris, Ann, 1956–
QM561.E635 1996
611′.0187′072–dc20 96–13077 CIP

ISBN 0 521 55023 8 hardback
ISBN 0 521 55991 X paperback

SE

Contents

Contributors

Hsiao Chang Chan
Department of Physiology, Faculty of Medicine, The Chinese University of Hong Kong, Shatin, New Territory, Hong Kong

Michele De Luca
Unit of Epithelial Biology and Biotechnology, Advanced Biotechnology Centre, Genoa, Italy

Nick Dove
Department of Clinical Biochemistry, Cambridge University, Addenbrooke's Hospital, Cambridge, UK

Robert Guy
Department of Clinical Biochemistry, Cambridge University, Addenbrooke's Hospital, Cambridge, UK

Angela Hague
CRC Colorectal Cancer Group, Department of Pathology and Microbiology, University of Bristol, School of Medical Sciences, Bristol, UK

Ann Harris
Paediatric Molecular Genetics, Institute of Molecular Medicine, Oxford University, John Radcliffe Hospital, Oxford, UK

Terence Kealey
Department of Clinical Biochemistry, Cambridge University, Addenbrooke's Hospital, Cambridge, UK

John R.W. Masters
Institute of Urology and Nephrology, Research Laboratories, London, UK

Chris Paraskeva

CRC Colorectal Cancer Group, Department of Pathology and Microbiology, University of Bristol, School of Medical Sciences, Bristol, UK

Graziella Pellegrini

Unit of Epithelial Biology and Biotechnology, Advanced Biotechnology Centre, Genoa, Italy

Michael Philpott

Department of Clinical Biochemistry, Cambridge University, Addenbrooke's Hospital, Cambridge, UK

Shirley E. Pullan

School of Biological Sciences, University of Manchester, Manchester, UK

Charles H. Streuli

School of Biological Sciences, University of Manchester, Manchester, UK

Stanley J. White

School of Biomedical Sciences, University of Sheffield, Sheffield, UK

Patrick Y.D. Wong

Department of Physiology, Faculty of Medicine, The Chinese University of Hong Kong, Shatin, New Territory, Hong Kong

Adrian S. Wolf

Developmental Biology Unit, Institute of Child Health, London, UK

James R. Yankaskas

Pulmonary and Critical Care Medicine, University of North Carolina, Chapel Hill, North Carolina, USA

Giovanna Zambruno

Laboratory of Molecular and Cellular Skin Biology, Istituto Dermopatico dell'Immacolata, Rome, Italy

Preface to the series

The series Handbooks in 'Practical Animal Cell culture' was born out of a wish to provide scientists turning to cell biology, to answer specific biological questions, the same scope as those turning to molecular biology as a tool. Look on any molecular cell biology laboratory's bookshelf and you will find one or more multivolume works that provide excellent recipe books for molecular techniques. Practical cell biology normally has a much lower profile, usually with a few more theoretical books on the cell types being studied in that laboratory.

The aim of this series, then, is to provide a multivolume, recipe-book-style approach to cell biology. Individuals may wish to acquire one or more volumes for personal use. Laboratories are likely to find the whole series a valuable addition to the 'in house' technique base.

There is no doubt that a competent molecular cell biologist will need 'green fingers' and patience to succeed in the culture of many primary cell types. However, with our increasing knowledge of the molecular explanation for many complex biological processes, the need to study differentiated cell lineages *in vitro* becomes ever more fundamental to many research programmes. Many of the more tedious elements in cell biology have become less onerous due to the commercial availability of most reagents. Further, the element of 'witchcraft' involved in success in culturing particular primary cells has diminished as more individuals are successful. The chapters in each volume of the series are written by experts in the culture of each cell type. The specific aim of the series is to share that technical expertise with others. We, the editors and authors, wish you every success in achieving this.

ANN HARRIS
Oxford, July 1995

Acknowledgements

First I wish to thank the busy scientists who put aside other pressing commitments to make the time to contribute a chapter to this volume.

Next thanks are due to the editorial team at the Cambridge University Press: to Robert Harrington who managed to persuade me that I should follow up my idea for this series, and to Tim Benton who convinced me that it was going to happen and then delivered the goods.

A.H.

Introduction

Ann Harris

Structure and function

Epithelial cells are organised in sheets of cells that make up the epithelia found at many sites in the body. The important function common to all epithelia is that of providing a barrier between the internal and external spaces of an organism. This includes external surfaces, such as skin, and internal surfaces that are exposed to the environment, such as airways and intestine. Epithelia are also found and carry out additional functions in the secretory parts of exocrine glands such as salivary glands and pancreas, and in the liver where the epithelial parenchymal cells, the hepatocytes, carry out the biochemical functions of liver.

Within different epithelia individual cells have specific functions relating to that epithelium: for example to secrete bicarbonate ions in the pancreatic duct, or to absorb sodium ions in the airway. All epithelia have a common set of characteristics, namely the ability to regulate movement of materials (ions, water, gases, nutrients and secretory products) in and out of the organ in which they are found. This may be through regulating permeability, exocytosis and endocytosis, or transport across the epithelium. These properties are dependent on the cells being polarised, that is they have distinct apical and basal surfaces with a different distribution of proteins at each. Thus, epithelial cells actively control passage of materials across the epithelium while the passive movement of materials by paracellular pathways is strictly limited.

The characteristic features of epithelia reflect their need to form an intact sheet in order to function. Individual epithelial cells are joined to each other through their interconnecting cytoskeletons. A large number of cell adhesion molecules are involved and specific areas of the cell membrane form intercellular junctions that (1) make the epithelium impermeable, (2) anchor cells to provide strength to the epithelium and (3) enable regulated movements of molecules between individual cells. In addition the epithelium as a whole is anchored to a basement membrane. All epithelia are in contact with a basement membrane that separates this surface cell layer from adjacent tissue. The basement membrane consists of an extracellular matrix of materials such as collagens, laminins, fibronectin and proteoglycans that are secreted both by the epithelial cells themselves and by the cells that underlie the membrane. Interactions with the extracellular matrix are crucial to differentiated function of epithelial cells and the absence of this matrix is likely to be responsible for lack of success in some epithelial cell culture systems.

Morphological characterisation of epithelia and epithelial cell types is widely used in their classification. This may be misleading because it does not necessarily reflect epithelial or cell functions; however, the different morphologies will be mentioned here as the terms are used in subsequent chapters. Squamous cells are flat, plate-like cells while the height of columnar cells is generally more than twice their diameter. Epithelia that are one cell layer thick are referred to as simple epithelia, while stratified epithelia consist of multiple layers of cells of which only the bottom layer is in contact with the extracellular matrix. Pseudostratified epithelium appears in cross-section to consist of multiple cell layers, but in fact it is unicellular in thickness and all cells are in contact with a common extracellular matrix. Transitional epithelia are restricted to the lining of the urinary tract, where their structure varies depending on the degree of stretching of the epithelium.

Embryological origin

With a few exceptions, including the epithelia lining the collecting ducts of the kidney, all epithelia are generated from the ectoderm or the endoderm of the developing embryo.

Stem cells

One interesting feature of epithelial cells is their ability continuously to regenerate, though the rate of cell division may be highly variable. Pancreatic duct epithelial cells replicate slowly, while fetal genital duct epithelial cells and adult epidermal keratinocytes replicate rapidly. One major investigative target for epithelial cell biologists has been the identification of stem cells in specific epithelia that provide this regenerative capacity. Each epithelium must contain proliferating cells that are able to replace those that become terminally differentiated and die. The potential of these stem cells to differentiate into distinct cell types found within each epithelium and the signals that govern these differentiation pathways are major areas of investigation that will be referred to in several of the chapters in this volume.

Importance in health and disease

Possibly as a direct result of the ability of epithelia to regenerate when damaged, they are common sites for the generation of tumours. The most common solid tumours in man – carcinomas of lung, breast, colon, bladder, pancreas and prostate – all derive from the malignant transformation of epithelial cells within those tissues. Hence these epithelial cells are the subject of extensive study. This has driven the establishment of many cell culture models for differentiated epithelium from a variety of sources. Many of these models have provided information on differentiation of the epithelium, though for many epithelia the precise nature of the stem cells remains to be elucidated. In addition epithelia are directly involved in many other human diseases, some of which are organ specific (for example colitis) while others affect multiple epithelia (such as the inherited disorder cystic fibrosis).

Aims of the volume

The following chapters contain methods for the culture of a number of primary epithelial cells types. The list of different cell types covered is not meant to be exhaustive. Instead culture systems that are reliable and have many direct applications have been chosen. The first chapter on airway epithelial cells presents one of the more technically demanding epithelial cell culture systems, yet it has wide-ranging applications in the study of diseases of the respiratory system. Chapter 2, on intestinal epithelial cells, provides a culture model that will be useful in addressing a wide range of questions on mechanisms of differentiation and tumorigenesis. Another epithelium in the digestive system, the pancreatic duct epithelium, is the subject of Chapter 3. Methods presented in this chapter could readily be modified for the culture of bile duct epithelial cells. Two chapters address the urogenital system: culture systems for renal and bladder epithelial cells are described in Chapter 4 and epididymal epithelial cells are covered in Chapter 5. Chapter 6 on mammary gland epithelial cells again provides an ideal culture sytem for investigating questions of differentiation and cancer. The final two chapters are on epithelial cells cultured from skin. Chapter 7, on the epidermal keratinocyte, contains several applications of this culture system. Chapter 8 presents culture systems for epithelial cells from skin glands and appendages.

Many of the chapters have techniques in common and it is likely that, faced with the task of establishing a culture system for any epithelial cell type, the reader will be able to develop an efficient methodology by drawing on the wealth of expertise presented here.

1

The airway epithelial cell

James R. Yankaskas

Introduction

The airways of the lungs conduct air between the environment and the alveolar spaces where oxygen and carbon dioxide exchange with the blood occurs. In this role, the airways warm and humidify inspired air, and defend the lungs from inhaled particles, toxins and microbes. The airways are normally lined by a continuous epithelium composed of cell types that provide specialised functions necessary for normal defences. Intercellular tight junctions connect adjacent cells along the apical cell borders and restrict permeability through the paracellular pathway. This physical barrier is a primary defence mechanism of the epithelia. The tight junctions also separate the cell membrane into apical and basolateral portions. Polarised distribution of proteins to distinct membranes is essential to the vectorial functions of airway epithelia.

Specialised defence mechanisms include the synthesis and secretion of mucus, transepithelial ion transport, mucociliary clearance, and interactions with the immune system. Airway epithelial cells produce mucins, antimicrobial molecules (e.g. lysozyme), anti-proteases (e.g. secretory leukopeptidase inhibitor) and other secreted molecules. Transepithelial ion transport (primarily of Na^+ and Cl^-) is responsible for salt and water balance in the layer of airway surface liquid lining the airways, and for mucus hydration. Epithelial cells release chemoattractant (e.g. LTB4) and other inflammatory molecules. Given their position as a primary interface with the environment, the airway epithelia have been proposed as the initiator and coordinator of immune responses to inhaled microbes and toxins.

A number of common lung diseases illustrate the importance of airway epithelial cell functions. The autosomal recessive disease cystic fibrosis (CF) is caused by mutations in the CFTR gene that lead to abnormal Na^+ and

Cl^- transport, chronic infection, respiratory morbidity and premature death. Abnormal mucociliary clearance in Kartagener's syndrome and other ciliary dyskinesia syndromes leads to chronic infection and airway obstruction. Exposure to tobacco smoke and some inhaled pollutants causes chronic bronchitis, emphysema and chronic obstructive lung disease. Cigarette smoke also causes bronchogenic cancer, a common and rapidly growing cause of mortality.

The specific roles of airway epithelial cells in normal physiology and disease are being elucidated, in part through evaluation of primary cell culture systems. This chapter describes isolation and culture methods for human airway surface epithelial cells. These cell culture methods can be modified to induce desired properties, such as maximal cell proliferation or differentiation into polarised epithelial preparations. Examples of such features are included in the text. The specific isolation and culture methods can be modified to achieve individual experimental goals.

Airway epithelial cell types

Airway generations and morphology

The extrathoracic airways begin in the nose and progress through the nasopharynx and larynx to the proximal trachea. The intrathoracic airways branch from the distal trachea through lobar, segmental and subsegmental bronchi to terminal bronchioles that lead to respiratory bronchioles and alveoli. There are about 23 bronchial generations from the trachea to the most distal bronchiole. The large increase in number of distal airways results in an increasing aggregate cross-sectional area in distal airways, resulting in slower airflows in these regions (Mercer *et al.*, 1994).

The airway structure and epithelial cell composition changes with generation number, and is best classified as bronchial (proximal) and bronchiolar (distal). Bronchi (including the trachea) have submucosal cartilage and range in diameter from 3 cm (trachea) to 3 mm (10th to 13th generation airways). Submucosal glands are present in the cartilaginous airways of humans, and variably present in other mammals. Bronchioles (14th to 23th generation) range in diameter from 3 mm to 0.4 mm, and end in respiratory bronchioles and alveolar acini. Gas exchange takes place in alveoli of the latter two structures.

The surface epithelium of the human nose and bronchi has pseudostratified columnar morphology that is primarily composed of ciliated, goblet and basal cells (Fig. 1.1). Ciliated and goblet cells have apical surfaces that face

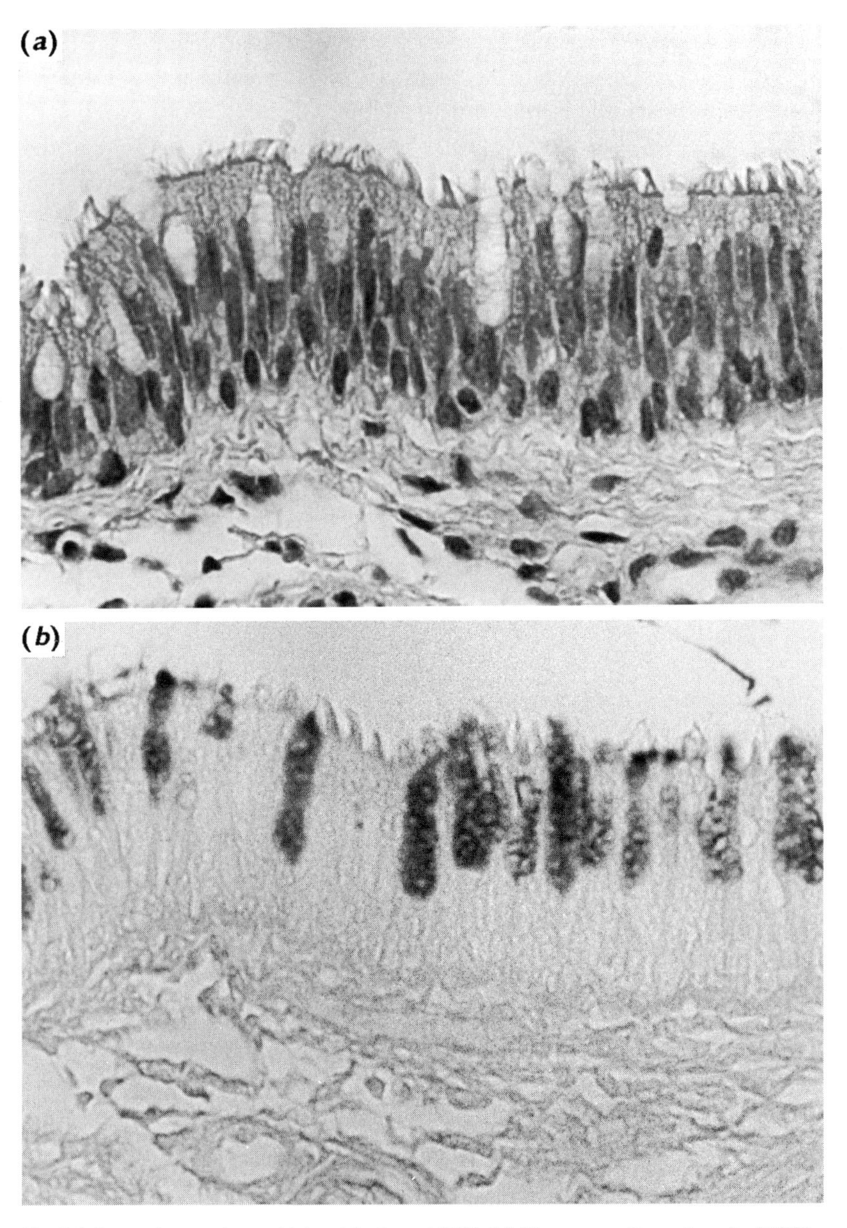

Fig. 1.1. Intact human bronchial epithelium, ×250. (*a*) Haematoxylin and eosin (H&E) stain. (*b*) Alcian blue–periodic acid Schiff (AB-PAS) stain. Basal cell nuclei are proximate to the basement membrane. Ciliated and goblet cells face the airway lumen. Goblet cell granules stain with AB–PAS.

the airway lumen, central nuclei, and a small area of attachment to the basement membrane. Basal cells have a larger area of basement membrane attachment, do not reach the airway lumen, and provide intercellular attachment for adjoining ciliated and goblet cells. The basal cell nuclei are close to the basement membrane and cause the stratified appearance of transverse histological specimens, even though all cells contact the basement membrane.

Ciliated cells

Ciliated cells have 150–200 cilia protruding from the apical membrane. Individual cilia are 3–7 μm long, 0.3 μm in diameter, and are comprised of microtubules, dynein and other proteins that provide coordinated motility. The characteristic 9+2 doublet pattern of these proteins seen on transmission electron microscopy is identical to that of other ciliated cells and spermatozoa.

Goblet cells

Goblet cells synthesise and secrete high-molecular-weight glycoconjugates. These neutral glycoproteins are stored in cytoplasmic granules and released by exocytosis when stimulated by purinergic agonists (e.g. extracellular ATP) and other undefined agonists. Goblet cells have a prominent perinuclear Golgi apparatus and numerous cytoplasmic granules that appear magenta with alcian blue–periodic acid Schiff (AB–PAS) stain.

Basal cells

Basal cells are smaller than ciliated and goblet cells, and are recognised by their characteristic location adjacent to the basement membrane, few cytoplasmic organelles, and a low cytoplasm/nucleus ratio. Human basal cells express cytokeratin 14, and are recognised by commercially available monoclonal antibodies specific to this intermediate filament.

Non-ciliated bronchiolar (Clara) cells

Bronchioles have a simple columnar morphology that is normally composed of ciliated cells and non-ciliated bronchiolar cells, commonly called Clara cells. Clara cells are taller than adjoining ciliated cells, bulge into the airway lumen, and have abundant smooth endoplasmic reticulum, consistent with their secretory function. Clara cells synthesise surfactant-related proteins A and B and a characteristic 10 kDa secretory protein (CCSP, or CC10). They are recognised by their unique morphology and by antibodies to CCSP.

Goblet, basal and Clara cells incorporate [^{3}H]thymidine in animal experiments and are capable of replication. Ciliated cells appear to be terminally differentiated. The exact lines of cell lineage have not been definitively established, and pluripotent stem cells for human airway epithelium have not been identified or isolated.

Other cell types

Intermediate cells are surface airway epithelial cells that do not express the specific identifying characteristics of any of the major cell types. They are typically located slightly above the basal cells in histological sections. In isolated cell preparations or cultures they may represent goblet cells that have degranulated or cells that have lost their differentiated characteristics during culture.

Surface serous cells are secretory cells that synthesise anti-inflammatory proteins such as lysozyme and secretory leukopeptidase inhibitor (SLPI). Their cytoplasmic granules appear blue in AB–PAS stained sections.

Submucosal glands develop in cartilaginous airways by invagination of the surface epithelium, and are comprised of ciliated and collecting ducts, and mucous and serous tubules (Fig. 1.2). The ciliated duct is lined by ciliated, goblet and basal cells similar to the surface epithelium. The collecting duct is lined by tall columnar cells that have numerous mitochondria and lack cilia and secretory granules. They may function to modify the fluid composition of glandular secretions. Mucous and serous cells have morphological and functional properties similar to goblet and surface serous cells. Submucosal gland serous cells also synthesise cystic fibrosis transmembrane conductance regulator (CFTR) protein, and may be responsible for fluid secretion necessary to flush mucous cell products onto the airway lumen. Isolation and culture methods similar to those described here have been developed for submucosal gland cells (Yamaya *et al.*, 1991).

Cell identification

Ciliated and other epithelial cells can be identified in histology sections by specific morphological criteria listed in this chapter. These features are variably retained in disaggregated and cultured cells, and alternative markers are being developed. Some cytokeratins (CK) are expressed in certain cell types, for example CK14 in basal cells and CK18 in ciliated and goblet cells.

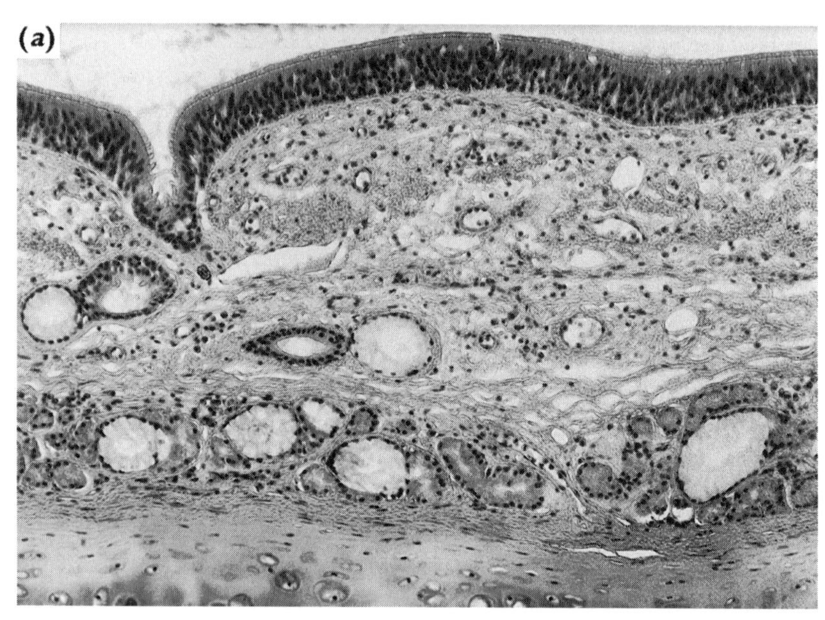

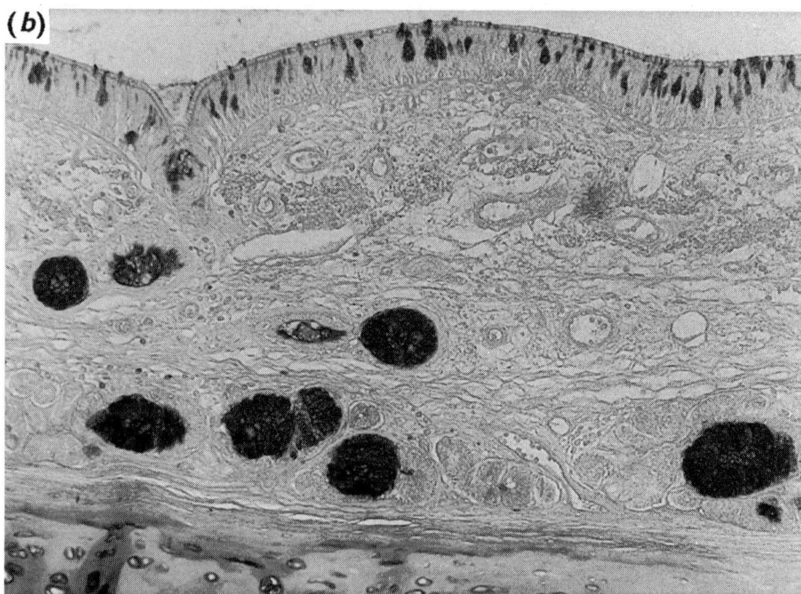

Fig. 1.2. Human bronchial submucosal gland, ×60. (*a*) H&E stain. (*b*) AB–PAS stain. Gland orifice appears as an epithelial invagination (left). Ciliated and collecting ducts near the orifice contain AB–PAS staining secretions. Mucous tubules are pale in H&E sections and stain with AB–PAS. Serous tubules have darker cytoplasm (H&E), minimal AB–PAS staining, and are situated peripheral to the mucous tubules.

Specific monoclonal antibodies for these intermediate filaments are available from several commercial producers, but reactivity requires cell fixation. Monoclonal antibodies have been raised against purified human mucins (Emery *et al.*, 1995) and screened by reactivity with excised human airways or cultured bronchial epithelial cells. The antibodies identify mucous or serous cells in the surface epithelium or submucosal glands, or combinations of these cells. Antibodies against cell surface epitopes recognise viable cells, and may be used for purification of individual cell types. Individual cell types from animal airways have been purified on the basis of cell surface markers. For example, rat tracheal basal cells labelled with the *Griffonia simplicifolia* I lectin were selected with a fluorescence-activated cell sorter (FACS) and established in primary culture (Randell *et al.*, 1991). Comparable purification of individual human airway epithelial cell types is possible, but has not been reported.

Specific applications

Cultured airway epithelial cells can be used for many experimental techniques. Recent examples from the field of CF research illustrate the breadth of this capability. Molecular genetics applications have included isolation of messenger RNA for the identification of the CFTR gene. Biochemical studies demonstrated that the increased sulphation of high-molecular-weight glycoconjugates produced by CF cells is an intrinsic property, rather than secondary to infection or inflammation. Ion transport studies of differentiated cultures, employing voltage-clamp methods and ion-selective double-barrelled microelectrodes, and of single cells by means of patch-clamp techniques, have elucidated the ion channel and regulatory abnormalities caused by the mutated CF gene. These techniques have also identified and characterised an alternative apical membrane Cl^- conductance in CF cells. Pharmacological studies of agonist concentration dependence and antagonists have identified the novel P2U purinergic receptor, which activates the alternative Cl^- conductance. Levels of the intracellular regulators Ca^{2+} and inositol phosphates have been evaluated using fluorescent dyes and high performance liquid chromatography (HPLC), respectively. Bacterial binding and epithelial interactions with inflammatory cells have been evaluated in differentiated cultures. Finally, the basis for pharmacological (aerosolised amiloride and UTP) and gene replacement (using viral and conjugated vectors) therapy for CF has been developed with primary and immortalised airway epithelial cells.

Sources of human cells

Nasal and bronchial tissues are excellent sources of proximal airway epithelial cells for several reasons. First, the epithelia lining these airways have similar morphology, cell type distribution and physiological functions. Second, nasal polyps and turbinates are frequently excised to relieve nasal obstruction, while grossly normal airway segments are often removed during lobectomy or pneumonectomy to treat localised tumours and other lung diseases. Third, significant amounts of epithelial tissue can be obtained from such specimens without any additional risk or discomfort to the patient. Bronchiolar samples can be obtained from distal lung specimens. Autopsy tissues can provide larger samples (including the trachea and mainstem bronchi), but the tissue may be damaged because of delayed autopsy and underlying disease processes such as infection and acidosis. Elective biopsies provide fresh cells, but are generally smaller than surgical samples, require patient cooperation, and may entail minor discomfort.

Research use of human tissues requires specific review and approval (by the Institutional Review Board (IRB) for the protection of rights of human subjects in the United States and by the local ethics committee in the United Kingdom). The research use of excess surgical and autopsy tissues may be authorised by the Operative or Autopsy Authorisation, or a separate permit may be required. The need for accurate pathological assessment takes precedence over most research uses.

Surgical specimens

Procurement of tissues removed for standard surgical indications requires the cooperation of surgical and pathology personnel. Following appropriate approval, a system must established to identify operations likely to produce usable tissues. The pathologist must identify tissue areas that are not required for pathological diagnosis. A skilled pathologist, surgeon or pulmonary physician should be on hand to dissect and initially process bronchial samples. Bronchioles can be obtained from portions of peripheral lung, typically cut into portions 1–2 cm thick.

Autopsy specimens

Tracheas, bronchi and bronchioles are obtained from autopsy samples by analogous techniques. Epithelial cell viability is best in tissues obtained within 3–6 h of death, and from patients who did not have airway infections, fluid overload, acidosis, or aspiration of gastric contents prior to death.

Research biopsies

Research airway epithelial biopsies can be obtained from the nose, trachea and large bronchi using brushes, plastic curettes or biopsy forceps (Bridges *et al.*, 1991; Kelsen *et al.*, 1992). Lower airway biopsies require bronchoscopy, which is safe when performed by experienced operators in appropriate patients. Brush and curette biopsies and forceps biopsy of lower airways cause minimal pain, but nasal forceps biopsy requires anaesthesia. Local anaesthetics (e.g. lidocaine) may cause a dose- and time-dependent decrease in cell viability.

Methods of tissue transportation and cell isolation

Epithelial cells in dissected tissue specimens remain viable for 1–3 days when kept in 4 °C physiological solutions. Excess blood and mucus is removed by rinsing the samples in sterile 0.9% saline, lactated Ringer solution or tissue culture media. For transportation, the tissues are placed in a 5- to 10-fold volume of Joklik's Modified Minimum Essential Medium (JMEM; Gibco 22300), Dulbecco's Modified Eagle Medium (DMEM; Gibco 23700), or comparable culture medium. Antibiotics (penicillin 100 U/ml; streptomycin 100 μg/ml; and gentamicin 40 μg/ml) should be added to kill bacteria that colonise the upper respiratory tract. An antifungal agent (amphotericin B, 2.5 μg/ml) is required for bronchiectatic lower airways from CF patients (but not for nasal tissues). The sample containers can be placed in Styrofoam shipping containers containing wet ice or reusable ice packs and transported to a research laboratory by overnight ground or air carrier.

Epithelial cells can be isolated from airway tissues by physical and/or biochemical dissociation, or can be allowed to migrate from the edges of epithelial tissues as explant cultures. Dissociation procedures can produce single cells, cell clusters or intact epithelial sheets, each having relative advantages. For example, single cells are required for cell sorting on the basis of surface markers, but require more stringent isolation techniques that may decrease cell yield. Intact sheets of epithelial cells retain cell–cell contacts, but may be difficult to orient on culture substrates. The overall experimental goals should be used to select isolation techniques that optimise the quantity, cell type and differentiation state of the experimental preparations.

Enzyme dispersion

Enzyme dispersion techniques (Yankaskas *et al.*, 1985) separate cells from the submucosal tissue by disrupting cell–substrate and cell–cell connections with enzymes and divalent cation chelating agents. Specific enzymes and disaggregation conditions are selected on the bases of cell yield, viability and selectivity for epithelial rather than mesenchymal cells. Dispersion techniques are particularly useful for small and irregular tissues that make physical methods difficult, and to provide large cell numbers. Additional physical methods such as scraping and brushing can reduce enzyme exposure but may damage cells. Proliferating and differentiated cultures can be established from single cell suspensions or from isolates that contain clusters of cells. Proliferation rate and time to reach maximal differentiation are influenced by seeding density, and may be improved by the presence of such clusters.

Protease 14 (Sigma P5147) is a non-specific protease that has been widely used to disaggregate airway epithelia from human and other mammalian tissues. The conditions in the following protocol have been optimised for epithelial cell yield and viability, although additional procedures (e.g. differential attachment or growth in serum-free medium) are required to eliminate fibroblasts. The protease incubation time can be extended to 48–72 h to modestly increase cell yield, but with lower cell viability and a reduced proportion of ciliated cells (Fig. 1.3).

Protease disaggregation

Materials

Joklik's Modified (calcium-free) Minimum Essential Medium (JMEM; GIBCO 22300), plus 15 mM HEPES (Sigma H-3375) and 24 mM $NaHCO_3$ (Sigma S-5761)

Ham's F-12 medium (GIBCO 21700) plus 15 mM HEPES and 24 mM $NaHCO_3$

Fetal bovine serum (FBS; mycoplasma, virus, bacteriophage and endotoxin tested; GIBCO 16000)

Protease/DNase: 1.0 mg/ml protease 14 (Sigma P-5147) plus 10 μg/ml DNase (Sigma DN-25) in JMEM (can prepare 10× concentrated protease/DNase and store in aliquots at −20 °C for 1–2 months)

Protocol

1 Remove tissue from transport medium and rinse with fresh JMEM.
2 Place tissue in a screw-capped sterile centrifuge tube containing 5-fold volume of protease/DNase. Mix thoroughly and incubate at 4 °C for

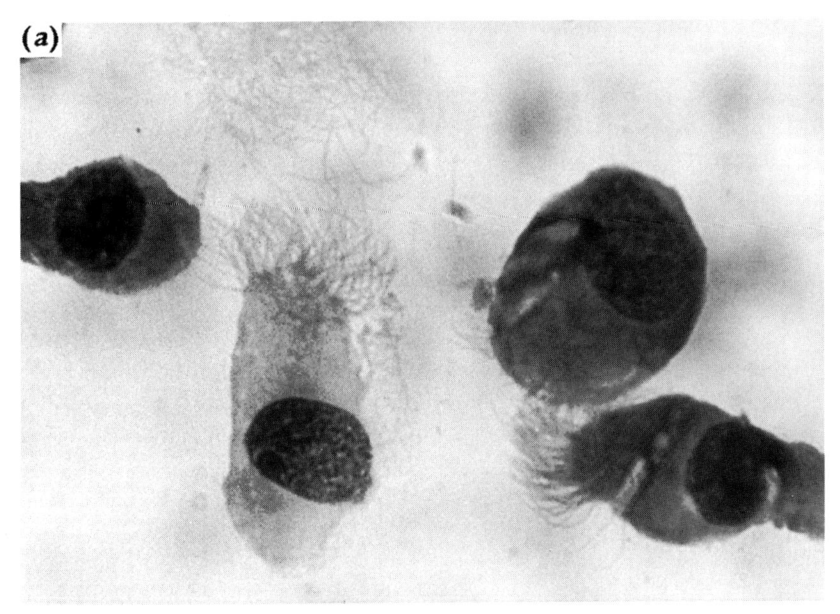

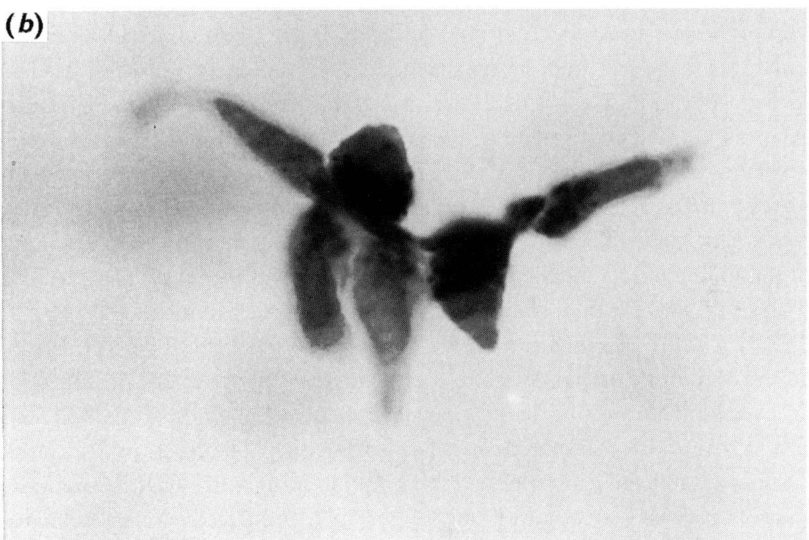

Fig. 1.3. Centrifuged primary cell isolates. (*a*). Ciliated cells, a goblet cell with eccentric nucleus and abundant cytoplasm, and a smaller basal cell. ×620, modified Wright stain. (*b*) Several goblet cells with eccentric nuclei and stained cytoplasmic granules. ×125, AB–PAS stain.

16–72 h. Mixing on a rocking platform at 10–20 cycles/min increases cell yield by ~80%.
3 Agitate the tube vigorously and remove tissue using sterile forceps. Loosely adherent cells can be removed from smooth surfaces of larger tissues (e.g. nasal turbinates with bone) by gently scraping the surface with a sterile scalpel. Add scraped cells to protease/DNase solution.
4 Add 10% FBS (v/v) to neutralise the protease.
5 Pellet cells (5 min at 500 *g*).
6 Resuspend the cells in JMEM+10% FBS. Pellet cells as above.
7 Resuspend the pellet in Ham's F-12 at about 1×10^6 cells/ml. (Repeated pipetting and/or incubation in DNAse (10 μg/ml) or dithiothreitol (Sigma D-9779, 0.5 mg/ml) for 10–15 min at 4 °C may be necessary to disrupt cell clumps.)
8 Count cells and plate as described below.

Explant

Epithelial cells at the edge of a small piece of tissue can migrate centrifugally and proliferate in explant cultures (Jefferson *et al.*, 1990). The cells form an enlarging plaque around the original tissue, with the peripheral cells undergoing replication. Cells on and near the explant tend to retain their columnar shape and cilia, probably due to interaction with submucosal tissue. Explant cultures can give rise to a large quantity of epithelial cells from relatively small tissue samples, but overall proliferation rates are relatively slow.

Explant cultures are prepared by sharply excising the surface epithelium, and mincing it into small pieces (about 1 mm^3). The pieces are placed about 1 cm apart on a plastic tissue culture dish in a minimal volume of Ham's F-12, DMEM, or F-12+DMEM media containing 10% FBS. The quantity of medium must be small enough to permit attachment to the dish (tissues float in excess medium), but desiccation must be avoided. Eight to ten explants can be plated in a 25 cm^2 flask, a bead of medium placed around the flask periphery, and the cap tightened after equilibrating with CO_2 to minimise evaporative losses. Medium is added sparingly, and after 2–4 days a normal quantity of medium (0.2 ml/cm^2) is added.

Intact epithelial sheets

Sheets of airway epithelial cells that retain the *in vivo* morphology can be cultured by a modified explant technique (Davis *et al.*, 1992). The epithelium is separated from the submucosal tissue by injecting submucosal collagenase

(100 Mandl U/ml) and dispase (grade II, 1 U/ml, both from Boehringer-Mannheim), incubation (37 °C, 30 min) and gentle scraping. The sheets are transferred to a nitrocellulose substrate, incubated overnight, then submerged in factor-supplemented medium. These cultures have minimal proliferation, and retain the native morphology, cell–cell interactions, and ciliary and goblet cell secretory functions. They are particularly useful for analysis of goblet cell secretion and regulation of ciliary beating.

Establishment and maintenance of cultures

Cell culture conditions should be selected and optimised to meet specific experimental goals. For example, biochemical assays may require cultures on plastic dishes in proliferation-stimulating medium; patch-clamp analyses of individual cells may require efficient attachment coatings; and studies of goblet cell secretion may require permeable substrates, an air interface, and retinoids in the medium. A large selection of attachment factors, medium additives and substrate designs has provided tremendous flexibility to the cell culturist. This section describes culture conditions that promote cell proliferation or polarised ion transport differentiation, respectively, and lists some alternatives that can be used to achieve other experimental goals.

Proliferation cultures

Maximal increases in cell number can be achieved by optimising attachment of primary isolates, and selecting growth factors that maximise proliferation rates (Wu *et al.*, 1985). In our laboratory, attachment efficiency (10–30 % with a seeding density of 3×10^4 cells/cm^2) is equivalent on standard tissue culture plastic and with collagen, fibronectin and/or albumin coatings. Serum-free F-12 medium is supplemented with seven growth factors (F-12+7X, Table 1.1) at concentrations selected to maximise proliferation. Medium is changed the day after plating, and every 2–3 days henceforth. The cultures are kept in a humidified incubator at 37 °C under 5% CO_2/95% air. Fibroblast growth is inhibited by the protease/DNase dissociation technique and use of serum-free media. Cell doubling time in these conditions is about 24 h. Proliferation rate is cell density dependent, and significantly impaired at confluencies <20%. Cells are generally passaged at 70–90% confluence, or used for experiments when they reach confluence, 5–14 days after plating.

Table 1.1 Growth factor preparation and storage: F-12+7× media supplements

Factor	Source[a]	Stock preparation	Stock cone	Storage	Expires	Final conc. in F-12+7×
Insulin (INS)	CR 40205	20 mg/4 ml H_2O	5 mg/ml	−20 °C	3 months	10 μg/ml
Hydrocortisone (HC)	CR 40203	(1) 50 mg/13.8 ml 70% EtOH	10^{-2}M	−20 °C	1 year	
		(2) dilute 1:10 in PBS	10^{-3}M	−20 °C	3 months	10^{-6} M
Endothelial cell growth supplement (ECGS)	CR 40006	15.0 mg/4 ml H_2O	3.75 mg/ml	−20 °C	1 month	3.75 μg/ml
Epidermal growth factor (EGF)	CR 40001	100 μg/4 ml H_2O	25 μg/ml	−20 °C	3 months	25 ng/ml
Triiodothyronine (T_3)	Sigma T6397	(1) 20 mg/5 ml 0.2 M NaOH				
		(2) add 5 ml H_2O	3 mM[b]	−20 °C	1 year	
		(3) dilute 1:100 in H_2O	3×10^{-5} M	−20 °C	3 months	3×10^{-8} M
Transferrin (T_f)	CR 40204	10 mg/4 ml H_2O	100 μg/ml	−20 °C	3 months	5 μg/ml
Cholera toxin (CT)	Sigma C3012	0.5 mg/4 ml H_2O	100 μg/ml	4 °C	1 year	10 ng/ml

Notes:

Vials of lyophilised powder stored at 4 °C.

Stocks do *not* need filtering in preparation if handled with sterile technique.

[a] CR, Collaborative Research, Inc., Two Oak Park, Bedford, MA 01730, USA. Telephone: ++1(617) 275-0004. Sigma, Sigma Chemical Company/PO Box 14608, St Louis, MO 63178, USA. Telephone: ++1(800)325-3010; foreign, ++1(314) 771-5750.

[b] Store 0.5 aliquots at −20 °C.

Differentiated cultures

Expression of some differentiated epithelial cell functions depends on formation of a confluent culture and cell polarisation. Such differentiation is supported by culturing on permeable supports, plating at high cell density, exposure of the apical cell surfaces to humidified air, and use of differentiation inducing media. The apical membrane, basolateral membrane and intracellular properties responsible for normal and cystic fibrosis Na^+ and Cl^- transport have been elucidated with such cultures (Yankaskas *et al.*, 1985; Willumsen *et al.*, 1989).

Airway epithelial cells isolated by protease disaggregation are seeded on collagen supports (Yankaskas *et al.*, 1985) or commercially available collagen-coated filters at a density of 3×10^5 cells/cm^2 in a minimal volume ($\sim$400 μl/cm^2) of F-12+7× medium. Sufficient submucosal media is added to maintain hydration and establish hydrostatic forces favouring flow towards the basolateral compartment. The preparation is placed in a humidified 37 °C incubator. Additional mucosal medium is gently added to the mucosal side the next day, and the medium is then changed every other day. When the cultures appear confluent by phase contrast microscopy, the cultures are changed to differentiation medium and the mucosal surface is left exposed to air. Differentiation medium consists of equal parts of F-12+7× and DMEM+2% FBS that had been conditioned by confluent 3T3 fibroblasts for 3 days (3T3-CM), plus retinoic acid (10^{-7} M).

In these conditions the cells achieve optical confluence 2–3 days after plating, and maximal transepithelial voltage and resistance develop 6–8 days after plating. With longer incubation (3–6 weeks), ciliogenesis occurs. The effects of the individual growth supplements listed in Table 1.1, and other culture conditions, on morphological and ion transport properties have been evaluated with dog (Van Scott *et al.*, 1988) and human tracheal (Yamaya *et al.*, 1992) epithelia. In brief, insulin and endothelial cell growth supplement are essential for cell survival. Epidermal growth factor contributes to a stratified morphology, and hydrated collagen gels, serum or serum substitutes, and exposure of the apical cell surface to air promote differentiation.

Morphological differentiation comparable to that of the original tissues can be attained by heterologous tracheal graft cultures (Engelhardt *et al.*, 1992). Epithelial cells obtained by protease disaggregation or removed from plastic dishes with trypsin/EDTA are inoculated in rat tracheas that have been denuded of epithelia by freeze–thaw cycles. The trachea is implanted in the subcutaneous space of an athymic immunotolerant mouse and incubated for 1–3 weeks. The airway lumen can remain closed, or be rinsed and exposed to air via plastic tubes that are attached to the trachea and tunnelled through a small skin incision. These preparations have been used for cell lineage, ion transport and gene therapy studies.

Longer-term culture and passage

The experimental uses of cultured cells can be amplified by increasing their number and by extending their lifespan. Primary human airway epithelial cells stop proliferating when the cells reach confluent density. Continued cell growth can be maintained by passing the cells to larger surface area plates (Technical Detail II). After three to six passages, cell attachment and proliferation decrease markedly, indicating the limit of cell survival in these culture conditions.

Cell proliferation and survival can be extended by co-culture with mouse 3T3 fibroblasts (American Type Culture Collection ATCC-CCL-92) that have been cultured in DMEM+10% FBS, then lethally irradiated (4000 rads) (Jefferson *et al.*, 1990). The fibroblasts are plated at 20 000–30 000 cells/cm^2 prior to or concomitantly with the epithelia. The fibroblasts die 5–7 days after plating, and can be removed by brief (1–3 min) trypsin/EDTA exposure. Alternatively, airway epithelial cells can be immortalised with the SV40 and human papilloma viral genes (Yankaskas *et al.*, 1993). Such cell lines have increased proliferation potential, and variably retain the phenotypic features of differentiated epithelial cells.

Passaging

Materials

Trypsin/EDTA: 0.1% trypsin type III (Sigma T0646)+1 mM EDTA (Sigma E6635) in phosphate-buffered saline (PBS)

Soybean trypsin inhibitor (STI, Sigma T9128) 1 mg/ml in F-12

Protocol

1 Aspirate medium and rinse cells with sterile PBS.
2 Add trypsin/EDTA, about 50 μl/cm^2, just enough to cover all cells.
3 Incubate at 37 °C.
4 Examine by phase microscopy every 2–3 min to assess cell detachment. Detach rounded cells by tapping the flask.
5 When >90% of cells are detached, add an equal volume of STI, to inhibit the trypsin.
6 Collect and pellet cells (500 *g*, 5 min).

7 Resuspend the cells at about 1×10^6/ml of plating medium. Count an aliquot in a haemocytometer.
8 Plate on fresh tissue culture plates at 1:3 or 1:4 dilution.

Cryopreservation

Cultured cells can be frozen and stored for months to years by the addition of cryopreservative agents that inhibit the formation of ice crystals. Primary airway epithelial cells are less tolerant of freezing than cell lines, and best results have been achieved by culturing on plastic plates for 4–6 days before trypsinising and freezing. The following protocols produce thawed cells with 80–90% viability.

Cryopreservation and thawing

Materials

2× freeze solution: F-12 medium+10% dimethylsulphoxide (DMSO)+ 10% FBS
1.8 ml cryovials (Nunc, Inc., 377267, Naperville, IL)

Protocol for freezing

1 Trypsinise cells and resuspend in culture medium at a density of $4–6\times10^6$ cells/ml. Place on crushed ice.
2 Add an equal volume of 2× freeze solution to the cell suspension drop by drop over 5 min. Mix by shaking intermittently.
3 Aliquot $2–3\times10^6$ cells (1 ml) to labelled cryovials.
4 Put vial(s) in an insulated Styrofoam box and place in a −70 °C freezer overnight to permit slow cooling. (Freezers with precisely controlled cooling rates are available, but expensive.)
5 Transfer vial to liquid nitrogen storage tank the next day.
6 After 2–5 days, thaw one vial and plate to confirm cell viability.

Protocol for thawing

1 Remove cryovial from liquid nitrogen and thaw in a beaker of 37 °C water, keeping the cap dry.
2 As soon as cells have thawed, add 1 ml of prewarmed plating medium to the cryovial and gently transfer the cells to an empty 15 ml tube. Add 10 ml prewarmed plating medium dropwise over 1–2 min to dilute the DMSO.
3 Pellet at 500 *g* for 5 min.
4 Resuspend cells in appropriate plating medium and plate at relatively high ($\sim8\times10^4$ cells/cm^2) seeding density.

Additional features

The methods described in this chapter are the basic techniques for isolation and culture of epithelial cells from human airways. The experimental utility of such cultures can be expanded by adaptation of these and other specific techniques. For example, individual cell types can be purified on the basis of physical properties (e.g. by centrifugal elutriation). Alternatively, cells can be labelled with specific primary antibodies and purified with secondary antibody-coupled magnetic beads or with fluorescent secondary antibodies and a fluorescence-activated cell sorter (FACS).

The phenotypic properties expressed by cultured epithelial cells depend on the substrate, composition of the medium and physical characteristics of the culture system. These conditions can be varied to meet specific scientific goals. For example, proliferation conditions are useful for isolation of constitutively produced cellular proteins, whereas differentiated cultures are required to evaluate vectorial ion transport or the regulation of ciliary beating. The interactions of epithelial cells with fibroblasts, inflammatory cells, toxins and soluble biological molecules can be evaluated by co-culture systems or direct exposures. The versatility of these culture systems is limited only by the ingenuity of the investigator.

Troubleshooting

The protocols described here have been adapted for use in laboratories around the world. Minor modifications are often necessary to optimise cultures for a specific experimental goal. Some approaches to dealing with common problems encountered in establishing primary airway epithelial cell culture capability are summarised below:

The viability of epithelial cells may deteriorate during tissue procurement and initial processing, transportation and cell isolation. Cells that stain with trypan blue (0.1% in PBS, 4 min) have lost membrane integrity. Evaluation of cell viability at sequential points in the isolation process can be used to identify and modify the injurious procedures.

Cell attachment to the substrate depends on viability, expression of membrane attachment molecules, culture substrate, calcium and physical contact with the substrate. Evaluation of alternative substrates in the presence of low (0.3 mM) or high (2 mM) Ca^{2+} or with 5% FBS in the attachment medium is a useful strategy. Replicate experiments with from three to six tissue samples may be required because of human tissue variability.

Cell proliferation can be influenced by variability in growth factors.

Comparing growth of cells in supplemented media obtained from different commercial sources or from an established research laboratory may focus the search for individual growth factors that inhibit cell proliferation. A large variety of commercially prepared permeable substrates is available, and a similar comparison of alternatives is often productive.

References

Bridges, M.A., Walker, D.C. & Davidson, A.G.F. (1991). Cystic fibrosis and control nasal epithelial cells harvested by a brushing procedure (letter to the Editor). *In Vitro Cell. Dev. Biol.*, **27A**, 684–6.

Davis, C.W., Dowell, M.L., Lethem, M. & Van Scott, M. (1992). Goblet cell degranulation in isolated canine tracheal epithelium: response to exogenous ATP, ADP, and adenosine. *Am. J. Physiol.*, **262**, C1313–23.

Emery N., Place, G.A., Dodd, S., Lhermitte, M., David, G., Lamblin, G., Perini, J.-M., Page, A.M., Hall, R.L. & Roussel, P. (1995). Mucous and serous secretions of human bronchial epithelial cells in secondary culture. *Am. J. Resp. Cell Mol. Biol.*, **12**, 130–41.

Engelhardt, J.F., Yankaskas, J.R. & Wilson, J.M. (1992). *In vivo* retroviral gene transfer into human bronchial epithelia of xenografts. *J. Clin. Invest.*, **90**, 2598–607.

Jefferson D.M., Valentich, J.D., Marini, F.C., Grubman, S.A., Iannuzzi, M.C., Dorkin, H.L., Li, M., Klinger, K.W. & Welsh, M.J. (1990). Expression of normal and cystic fibrosis phenotypes by continuous airway epithelial cell lines. *Am. J. Physiol.*, **259**, L496–505.

Kelsen S.G., Mardini, I.A., Zhu, S., Benovic, J.L. & Higgins, N.C. (1992). A technique to harvest viable tracheobronchial epithelial cells from living human donors. *Am. J. Respir. Cell Mol. Biol.*, **7**, 66–72.

Mercer R.R., Russell, M.L., Roggli, V.L. & Crapo, J.D. (1994). Cell number and distribution in human and rat airways. *Am. J. Respir. Cell Mol. Biol.*, **10**, 613–24.

Randell S.H., Comment, C.E., Ramaekers, F.C.S. & Nettesheim, P. (1991). Properties of rat tracheal epithelial cells separated based on expression of cell surface alpha-galactosyl end groups. *Am. J. Respir. Cell Mol. Biol.*, **4**, 544–54.

Van Scott M.R., Lee, N.P., Yankaskas, J.R. & Boucher, R.C. (1988). Effect of hormones on growth and function of cultured canine tracheal epithelial cells. *Am. J. Physiol.*, **255**, C237–45.

Willumsen, N.J., Davis, C.W. & Boucher, R.C. (1989). Intracellular Cl^- activity and cellular Cl^- pathways in cultured human airway epithelium. *Am. J. Physiol.*, **256**, C1033–44.

Wu, R., Yankaskas, J., Cheng, E., Knowles, M.R. & Boucher, R. (1985). Growth and differentiation of human nasal epithelial cells in culture: serum-free, hormone-supplemented medium and proteoglycan synthesis. *Am. Rev. Respir. Dis.*, **132**, 311–20.

Yamaya, M., Finkbeiner, W.E. & Widdicombe, J.H. (1991). Ion transport by cultures of human tracheobronchial submucosal glands. *Am. J. Physiol.*, **261**, L485–90.

Yamaya M., Finkbeiner, W.E., Chun, S.Y. & Widdicombe, J.H. (1992). Differentiated structure and function of cultures from human tracheal epithelium. *Am. J. Physiol.*, **262**, L713–24.

Yankaskas J.R., Cotton, C.U., Knowles, M.R., Gatzy, J.T. & Boucher, R.C. (1985). Culture of human nasal epithelial cells on collagen matrix supports. *Am. Rev. Respir. Dis.*, **132**, 1281–7.

Yankaskas J.R., Haizlip, J.E., Conrad, M., Koral, D., Lazarowski, E., Paradiso, A.M., Rinehart, C.A. Jr, Sarkadi, B., Schlegel, R. & Boucher, R.C. (1993). Papilloma virus immortalised tracheal epithelial cells retain a well-differentiated phenotype. *Am. J. Physiol.*, **264**, C1219–30.

2

The intestinal epithelial cell

Angela Hague and Chris Paraskeva

Introduction

Colorectal carcinogenesis is an excellent example of the multistage nature of tumour progression. Most carcinomas are derived from premalignant adenomas (polyps) and the adenomas are derived from the normal colorectal epithelium. Cell lines from different stages of the adenoma to carcinoma sequence form a valuable resource for studies of the cellular and molecular biology of colorectal cancer. Transfection of specific genes can be used to define the function of these genes in the background of the colorectal tumour cell. The cultures are also used to determine the effects of drugs on cell behaviour and gene expression. Cell culture techniques which permit the growth and differentiation of colonic epithelium from normal, premalignant and malignant tissue are particularly valuable. We have shown that apoptosis can be measured in these cultures (Hague *et al.*, 1993); therefore cell lines can be used to assess the balance between proliferation, differentiation and cell death. The culture of normal colonic epithelium is, as yet, limited to short-term cultures of adult tissue or to growth of tissue of fetal origin. Normal adult colonocytes undergo apoptosis shortly after transfer into culture (unpublished observations).

Friedman *et al.* (1981) have described conditions for the primary culture of adenoma cells and the protocol described here is suitable for culture and establishment of cell strains and cell lines of adenoma cells as well as carcinoma cells (Paraskeva *et al.*, 1984, 1989). The fact that adenoma cells are able to survive long-term culture indicates that the acquisition of an apoptosis-resistant phenotype is an early stage in the development of a colorectal tumour. The successful establishment of cell strains/lines relies on the maintenance of cell–cell contacts and to achieve this passaging is carried out using dispase, which removes the cells as clumps rather than as single cells. The adenoma cell strains may senesce after a number of passages or may become immortal cell lines depending on the stage of tumour progression. Later-stage adenomas may also survive after trypsinisation to single cells (i.e. are clonogenic). Clonogenicity can therefore be used as a further marker of tumour progresssion. Normal colonocytes and early-stage adenomas do not survive dissociation to single cells. Bedi *et al.* (1995) have demonstrated that normal colonic epithelial cells undergo apoptosis if cell–cell contacts are broken and we have also observed this in the early stage non-clonogenic adenomas (unpublished observations). An important stage in colorectal carcinogenesis, therefore, is the evasion of the programmed cell death pathway(s) and this stage is separable from the acquisition of immortality.

Colorectal adenocarcinomas are relatively easy to establish as cell lines and several methods have been described for their culture (Leibovitz *et al.*, 1976; Brattain *et al.*, 1983; Paraskeva *et al.*, 1984; Kirkland & Bailey, 1986). Carcinomas give rise to immortal cell lines and are usually able to grow from single cells. However, for establishment of cell lines, maintenance of cell–cell contacts, as for the adenoma cultures, reduces the severity of the selection imposed. A similar method is therefore used for establishment of both adenoma and carcinoma cultures.

Characterisation and morphology

Epithelial cells have a classic cuboidal morphology which gives cultures the appearance of pavement-like sheets, but morphology is only a clue to the nature of the cells since, for example, endothelial cells at high density can have an epithelial-like morphology. Further characterisation is therefore necessary. A useful marker for epithelial cells is the staining of cytokeratin filaments using anti-keratin monoclonal antibodies (e.g. LE61 or Dako-LP34). In addition differentiation markers characteristic of the cell type can be used to assist in the confirmation of the epithelial nature of the cells. In colonic epithelium there are three differentiated cell types: mucus-secreting goblet cells, fluid-transporting enterocytes and endocrine cells. Electron microscopic examination can be used to demonstrate desmosomes, microvilli and mucins in the goblet-like cells. The mucin, which may be secreted into the culture medium, can be detected using histochemical stains (e.g. alcian blue–periodic acid Schiff) (Paraskeva *et al.*, 1984) or by the use of antibodies to intestinal mucins MUC2 and MUC3 (Jass & Roberton, 1994; Chang *et al.*, 1994). Enterocytic cells can give rise to domes, or hemicysts, as fluid transported vectorially accumulates between the cell monolayer and the culture vessel. The enterocytic cells also produce carcinoembryonic antigen

(CEA) (which can be detected using anti-CEA monoclonal antibodies) and alkaline phosphatase, the activity of which can be evaluated using an enzyme assay (Sigma, procedure 245). The cell adhesion molecule E-cadherin can also be used as a marker of differentiation in that if the cells do not express E-cadherin, they will be poorly differentiated (Pignatelli, 1993). However, as a cautionary note, cells may express non-functional E-cadherin. The following protocol produces colorectal adenocarcinoma cell lines in which the above differentiated phenotypes are expressed.

Carcinoma cells will normally form tumours on injection into athymic nude mice, whereas adenoma cells will not. It is possible to convert adenoma cells to a tumorigenic phenotype in an *in vitro* transformation assay (Williams *et al.*, 1990). The tumours formed in the mice have the same epithelial characteristics and differentiation pattern as the tumour from which the cell line was derived.

Source of cells

Colorectal carcinoma specimens are collected at surgery whereas adenomas can be collected either at endoscopy or at surgery (if a carcinoma is removed along with associated adenomas). Normal colonic mucosa can be obtained when surgery is performed for non-neoplastic intestinal diseases. Specimens must be placed directly in washing medium (see below) to prevent drying out. It is important that the specimen is collected promptly after removal from the patient, as extensive apoptosis occurs in the colonic crypts subsequent to resection (Hall *et al.*, 1994). The samples should be kept on ice for transportation to the laboratory. Note that gloves should be worn for handling human tissue.

Washing medium

Dulbecco's Modified Eagle's Medium (DMEM) supplemented with 5% fetal bovine serum (FBS), 2 mM glutamine, 200 U/ml penicillin, 200 μg/ml streptomycin and 50 μg/ml gentamicin.

Preparation of solutions to supplement medium

Glutamine solution

The final concentration is 2 mM. The stock is made up at 200 mM in tissue culture tested water. The stock solution is filter sterilised (using a 0.2 μm filter), aliquoted and stored at −20 °C.

Penicillin/streptomycin solution

The antibiotics can be dissolved together when preparing the stock solution. Penicillin is used at 100 U/ml and streptomycin at 100 μg/ml for *standard culture medium*. The stock solution should be made up with each of the antibiotics at 100× these working concentrations, although in *primary culture growth medium* and *washing medium*, the antibiotic concentration is twice that normally used for culture maintenance. The antibiotics are dissolved in tissue culture tested water and are filter sterilised. Aliquots are stored at −20 °C.

Methods of isolation

Materials
Sterile forceps and scalpels
Sterile petri dishes (glass preferable to plastic)
Sterile glass 10 ml pipettes

Solutions and reagents
Hyaluronidase solution
Collagenase solution
Washing medium
Primary culture growth medium
Collagen coated flasks
Swiss 3T3 feeders

Preparation of solutions

Hyaluronidase solution is used in the digestion of the tumour tissue. Hyaluronidase type 1-S (Sigma) is used at a working concentration of 100 U/ml. The stock solution is made up at 1000 U/ml in warmed DMEM supplemented with penicillin (100 U/ml) and streptomycin (100 μg/ml). The solution is filter sterilised and frozen as 2 ml aliquots.

Collagenase solution is used in the digestion of the tumour tissue along with hyaluronidase. Collagenase ClsIII (Worthington Biochemical Group; Lorne Diagnostics, UK) is used at a final concentration of 240 U/ml. The stock solution is made up at 480 U/ml in warmed DMEM supplemented with penicillin (100 U/ml) and streptomycin (100 μg/ml). The solution is filter sterilised and frozen as 10 ml aliquots.

Primary culture growth medium is DMEM supplemented with 20% FBS (batch selected), hydrocortisone sodium succinate (1 μg/ml) insulin (0.2 U/ml), glutamine (2 mM), penicillin (200 U/ml), streptomycin (200 μg/ml) and gentamicin (50 μg/ml) (Flow Laboratories).

Hydrocortisone 21-hemisuccinate (sodium salt; Sigma) is used to enhance the cell yield. The working concentration is 1 μg/ml. Stock solution is made up at 100 μg/ml by dissolving the hydrocortisone in warmed DMEM containing penicillin (100 U/ml) and streptomycin (100 μg/ml). The solution is filter sterilised and aliquoted into 5 ml amounts. Hydrocortisone may be stored at −20 °C, but must not be refrozen after thawing. Once thawed, any unused solution must be stored at 4 °C.

Preparation of collagen coated flasks

Materials

Sterile stirring bar
Sterile forceps
50 mg human placental collagen type VI (Sigma)
Sterile tissue culture tested water

Protocol

1 Sterilise collagen as powder by gamma irradiation (2.5 MRad, 25 kGy).
2 Add 0.05 ml glacial acetic acid to 50 ml warm sterile tissue culture tested water.
3 Add 50 mg collagen using sterile forceps.
4 Leave stirring at 37 °C for 2–3 h until completely dissolved.
5 Pipette into four universal containers and centrifuge at 1600 *g* for 3 min to remove impurities.
6 Transfer collagen supernatants to fresh universal containers and store at 4 °C.
7 To coat the flasks, keeping the collagen solution on ice, add about 4 ml to a flask, coat the bottom of the flask, pipette the collagen out and into the next flask, and so on. Aspirate off excess collagen solution. This will leave a thin coat of collagen on the bottom of the flasks.
8 Leave the flasks to air dry horizontally in the laminar flow hood with the lids off for 2–4 h.
9 Store the collagen coated flasks for up to 4 weeks at 4 °C.

Preparation of 3T3 feeder cells

3T3 cells are treated with mitomycin C at a concentration of 10 μg/ml for 2 h to prevent them from growing. An alternative method is to lethally irradiate the cells with 60 kGy (6 MRad) gamma irradiation. Although unable to grow, the 3T3 feeders attach to the substrate and produce growth factors when added to epithelial cell cultures.

Materials

Swiss 3T3 cells (American Tissue Type Culture Collection, no. CCL92)
Mitomycin C (Sigma)
Sterile phosphate-buffered saline (PBS) (autoclaved)
Sterile 0.1% trypsin in PBS (filter sterilised)

Gloves should be worn when handling mitomycin C.
Mitomycin C is light sensitive, therefore hood lights should be switched off.

Protocol

1 Grow Swiss 3T3 cells on 90 mm petri dishes in DMEM supplemented with 10% calf serum and 2 mM glutamine, plus antibiotics as for standard culture medium, until they are 24 h post-confluent.
2 Prepare mitomycin C as a 10 μg/ml stock solution in the following way. Dissolve 2 mg mitomycin C in 2 ml medium warmed to 37 °C and added by a syringe into the vial. Pierce the rubber stopper with another syringe needle to release the pressure before removing the solution from the vial. Make the volume up to 200 ml with warm medium.
3 Remove the medium from the 3T3 cells and replace with 10–12 ml of the mitomycin-C-containing medium. Incubate for 2 h (no less).
4 Carefully remove the mitomycin C and rinse the cells with sterile PBS (at 37 °C).
5 Wash the cells using 0.1% trypsin and then aspirate off most of the trypsin, leaving only a thin layer covering the cells.
6. Incubate for 5–10 min to allow the cells to detach.
7 Suspend cells in medium and centrifuge at 1600 *g* for 3 min.
8 Resuspend the pellet in fresh medium.
9 Repeat steps 7 and 8 a further three times to wash mitomycin C off the cells.
10 Count the cells using a counting chamber and make up the cell suspension to 1×10^6 cells/ml.
11 Store the prepared feeder cell suspension at 4 °C and use for up to 1 week.

Methods of isolation

The disaggregation technique used will depend on the histology and structure of the tumour specimen. Poorly differentiated tumours (which lack uniform arrangement into crypts or tubules) may only require physical cutting using scalpels, since such tumours have little glandular organisation and small clumps of tumour cells are released into the surrounding washing medium when cut. A well-differentiated tumour (with well-formed glandular structures) may not release small clumps of cells on cutting and in such cases enzyme digestion is necessary. An overnight digestion of tumour pieces approximately 1 mm^3 in size in collagenase and hyaluronidase also digests associated mesenchymal tissue away from the released epithelial organoids. Organoids are then cleaned by repeated pipetting and gravity sedimentation. Epithelial organoids will settle out, whereas the more diffuse mesenchymal tissue has a tendency to remain floating.

Preparation of colorectal adenocarcinoma tissue for culture

Protocol

1. Wash the tumour biopsy four times in washing medium.
2. Fix a small sample of the tumour specimen for histological diagnosis. This can subsequently be compared with the tumour histology reported by the hospital pathologists.
3. In a very small volume of washing medium, cut the remaining tissue using crossed scalpel blades until the pieces are 1 mm^3 or less. Cutting should be continued for at least 10 min. The pieces should be small enough to be pipetted easily with a 10 ml glass pipette. It is advisable to cut only 1 cm^3 of tissue at any one time, so that smaller pieces can be obtained.
4. Wash the tissue pieces four times by suspension followed by bench centrifugation (1600 *g* for 3 min). If, on cutting the tumour sample, small clumps of tumour cells were released, enzymatic digestion will not be necessary; therefore go to step 9. If in doubt, or if the tumour is a rare specimen, digest half the sample.
5. Resuspend the tissue pieces in 8 ml washing medium per 1 cm^3 tissue in a universal container.
6. To the 8 ml suspension, add 2 ml hyaluronidase solution and 10 ml collagenase solution.
7. Incubate the samples at 37 °C on a rotator, usually overnight (12–16 h).
8. Centrifuge at 1600 *g* for 3 min and remove the enzymes. If the sample

is very mucinous, this will also remove some of the mucin and make separation easier. As a result of the presence of mucin, care should be taken that the pellet is not lost on aspirating the supernatant. If necessary, it is better to remove most of the supernatant, wash and centrifuge down once more.

9 Resuspend the sample in washing medium and vigorously pipette up and down using a bulb pipette filler to clean mesenchymal tissue from the epithelial organoids. Allow the pieces to settle by gravity sedimentation (5–15 min).
10 Remove the supernatant and examine microscopically. It will contain many fibroblastic cells. If small clumps of epithelial cells are present, transfer to a separate tube and allow to settle. If gravity alone will not bring down the cell clumps, differential centrifugation (160 *g* for 1–2 min) can be used. (Retaining cells from the supernatant can have two uses: (i) it safeguards against discarding more dysplastic elements of the tumour, (ii) it can be a good source of fibroblast cells which can be used for molecular studies as a supply of the normal constitutional DNA.)
11 Repeat the gravity sedimentation, resuspension and pipetting four times for each tube.
12 Examine the settled organoid material microscopically. If organoids have been cleaned of fibrous material, centrifuge at 1600 *g* for 3 min; if not, repeat steps 10 and 11. Fig. 2.1 shows the appearance of cleaned organoids from an adenoma.
13 Remove the supernatant and resuspend in 4 ml primary culture growth medium per tube.
14 Transfer to collagen coated flasks, and to each T25 flask add 0.1-0.2 ml 3T3 feeders (1×10^5 to 2×10^5 cells).
15 Incubate at 37 °C in 5–10% CO_2 in air. (Note that the use of dry incubators significantly reduces contamination problems.)

Establishment and maintenance of cell lines from colorectal tumours

Once the organoid material has been put into culture, attachment should be seen after 24–48 h and epithelial cells will start to migrate out of the central organoid structure. Some epithelial organoids, however, may take up to 6 weeks to attach, and until all the epithelial organoids have adhered to the culture vessel, non-adherent material should be centrifuged, resuspended in fresh growth medium and replated when changing the medium. In some cases the cell population will continue to expand until the culture can be pas-

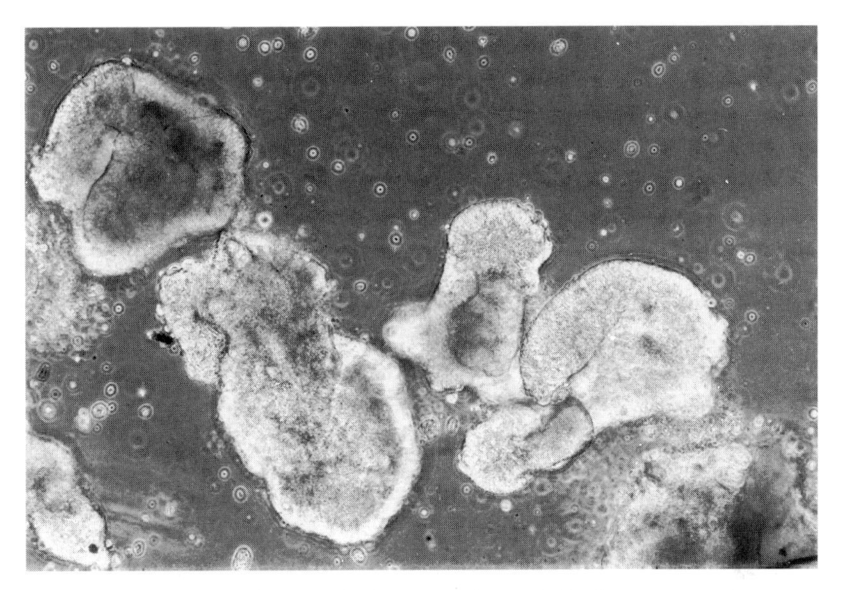

Fig. 2.1. Phase contrast photograph of adenoma 'organoids' after removal of fibroblasts by the pipetting and sedimentation technique (×80). Note the smooth edges of the tumour pieces.

saged; in other cases the culture may do well for a while but then regress and only a few colonies survive. It may be up to 3 months before passaging can be attempted. Fig. 2.2 shows the growth pattern of a colonic adenoma primary culture after 15 days in culture, demonstrating pavement-like epithelial morphology of cells growing out from the organoids. This culture was subsequently successfully passaged and a cell line established. After a few trial runs, it should be possible to obtain a 50–80% success rate of establishing colorectal tumour cell lines.

Colorectal adenocarcinoma cells are routinely grown on collagen type IV coated T25 tissue culture flasks in the presence of Swiss 3T3 feeder cells (at a density between 1×10^3 and 1×10^4 cells/cm^2) at 37 °C in an incubator under 5% CO_2 in air. The medium should be changed twice weekly. The 3T3 feeder cells must be maintained at 30–40% confluence, and this will normally entail replenishment at the rate of approximately 0.1 ml (10^5 cells) per week. The maintenance of 3T3 feeder support is important. It is necessary to maintain a high density, but if too many feeder cells are present this may prevent attachment of the epithelial organoids. For both adenoma and carcinoma cultures success can be achieved using 3T3 conditioned medium either as an alternative to or in addition to 3T3 cells. The use of conditioned medium can be a useful way of removing the 3T3 feeder cells from the

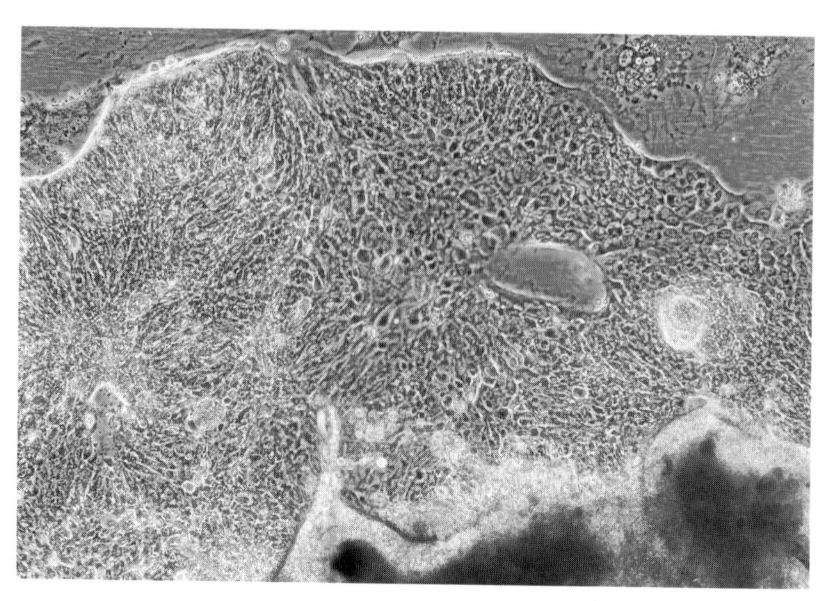

Fig. 2.2. Phase contrast photograph of a 15 day primary culture from a colonic adenoma (×85). Note the dense organoid structure at the bottom right has attached and cells are migrating out to give a sheet of pavement-like epithelium. The epithelial cells have a characteristic cuboidal shape.

cultures. This may be desirable for the production of pure samples of epithelial cells. Variants can be derived which can grow without any 3T3 feeder support, but in the early stages it is advisable to maintain a high cell density.

For the first 2 weeks of culture, the high antibiotic concentrations of the primary culture growth medium should be maintained. After this time, the standard culture medium may be used, i.e. Dulbecco's Modified Eagle Medium (DMEM) supplemented with 20% fetal bovine serum (FBS), hydrocortisone sodium succinate (1 μg/ml), insulin (0.2 U/ml), glutamine (2 mM), penicillin (100 U/ml) and streptomycin (100 μg/ml). The batch of FBS used can greatly affect the growth of colorectal adenocarcinoma cells. It is important to test any batch of serum before purchasing, using a recently established colorectal epithelial cell line – preferably one which exhibits a differentiated phenotype *in vitro*. If the aim is to grow primary cultures it is important not to use a well-established cell line for batch-testing serum.

A major obstacle in the successful establishment of a new cell line is the presence of contaminating fibroblasts. These should be removed from the culture as they can rapidly outgrow the epithelial cells. There are a number of ways of achieving this. Firstly, well-isolated fibroblastic colonies can be removed by scraping with a cell scraper or the plunger of a sterile syringe.

The flasks should then be washed several times to remove the fibroblasts. Alternatively the differential responses of the cells to the enzymes trypsin and dispase can be used selectively to remove fibroblasts or epithelial cells from the flask. Differential trypsinisation can be used to remove fibroblasts without the epithelial cells detaching. However, if growing benign adenomas, this technique should not be used, since adenoma cells are more sensitive to trypsin than are carcinoma cells and may be killed. Dispase can be used preferentially to remove the epithelium. This leaves many fibroblastic cells behind on the flask. It also gives the opportunity for cleaning the epithelial organoids further by vigorous pipetting or, in the case of severe fibroblast contamination, for further digestion in collagenase/hyaluronidase. The cells can be incubated on tissue culture plastic for 2–6 h which will permit preferential attachment of fibroblasts. The organoids can then be transferred to a fresh collagen coated flask. An additional method is to use a conjugate between an anti-Thy-1 antibody and the toxin ricin (Paraskeva *et al.*, 1985), which allows specific killing of fibroblasts, but unfortunately this conjugate is not yet commercially available. Each of the methods given here has been used with success, but in cases where there is close association between the epithelial and the mesenchymal cells it is often very difficult to eliminate the fibroblasts which grow out from underneath the epithelium. Time spent in cleaning the epithelial cell clumps during the preparation of the tissue before culture is therefore well invested.

Longer-term culture and passage

Although carcinoma cells can be passaged using trypsin, this may be selective. Passaging the cultures as cell clumps using dispase retains cell–cell contacts and permits cell differentiation. From primary culture, several 1:1 passages may be required before there are sufficient cells for a 1:2 split.

Dispase solution

Dispase Grade 1 (Boehringer Mannheim) is used at a working concentration of 2 U/ml. The dispase is dissolved in warmed medium containing 10% FBS, 2 mM glutamine, penicillin (100 U/ml) and streptomycin (100 μg/ml), filter sterilised and frozen as 5 ml aliquots. On thawing, dispase solution should be centrifuged before use (1600 *g* for 3 min) to remove any precipitate.

To passage using dispase, remove the culture medium and add 2.5 ml dispase into each T25 flask. Established cell lines should take 30 min to detach, but primary cultures may take 60 min or longer. Monolayers will detach as sheets; less confluent cultures will detach as clumps. Some cultures may need agitation to remove the cells from the flask. Add 7.5 ml standard growth medium to the flask to help wash off the cells and transfer to a universal container. Centrifuge at 1600 *g* for 3 min, remove the supernatant and resuspend the cell clumps in standard growth medium. Repeated pipetting up and down will break the cell clumps up a little and give a better distribution on the flask after seeding. Return the cell clumps to fresh collagen coated flasks and add 0.1 ml (10^5 cells) 3T3 feeders. The feeder cells are particularly important after passaging, so it is worth checking that the distribution of feeders is satisfactory after the cells have attached.

Subsequent to successful establishment of a cell line, carcinoma cell lines can usually be passaged using 0.1% trypsin/0.1% EDTA (dissolved in PBS to be isotonic to the cells). Some adenoma cell lines cannot be passaged as single cells (i.e. are non-clonogenic) and dispase should be used to remove the cells as clumps. Even for carcinoma cells, it is best to use dispase for the first few passages, as this treatment will be less severe. Cultures can also be grown on tissue culture plastic instead of collagen and feeder-independent cell lines can be derived. These culture conditions are useful for certain experiments, but the cultures will grow at a reduced rate and the differentiation pattern of the cells may be altered. Removal of collagen and 3T3 feeder cells results in a significant reduction in the number of mucin-secreting goblet cells and a corresponding increase in brush border columnar-like cells. However, colon cancer cell lines can be grown on a simple medium consisting of DMEM supplemented with 10% FBS and glutamine only, without a collagen substrate and without 3T3 feeders, and still retain the ability to produce some mucin glycoproteins.

Cryopreservation

For cryopreservation, cells are first removed from flasks using dispase, centrifuged and resuspended in standard culture medium containing 10% dimethylsulphoxide (DMSO). Sufficient cells should be added to each freezing ampoule to enable satisfactory recovery on thawing. For example, if the split ratio of a culture is 1:4 for passaging, then two ampoules can be frozen per confluent flask of cells. Cells should be frozen slowly above liquid nitrogen for at least 3 h or overnight, and subsequently placed in liquid nitrogen. Recover one ampoule of cells to check that the frozen cells are viable. To thaw, the ampoule should be warmed rapidly in a 37 °C water bath. Immediately after thawing, add the contents of the vial to 9 ml of medium and centrifuge. Remove the supernatant containing the DMSO (which is

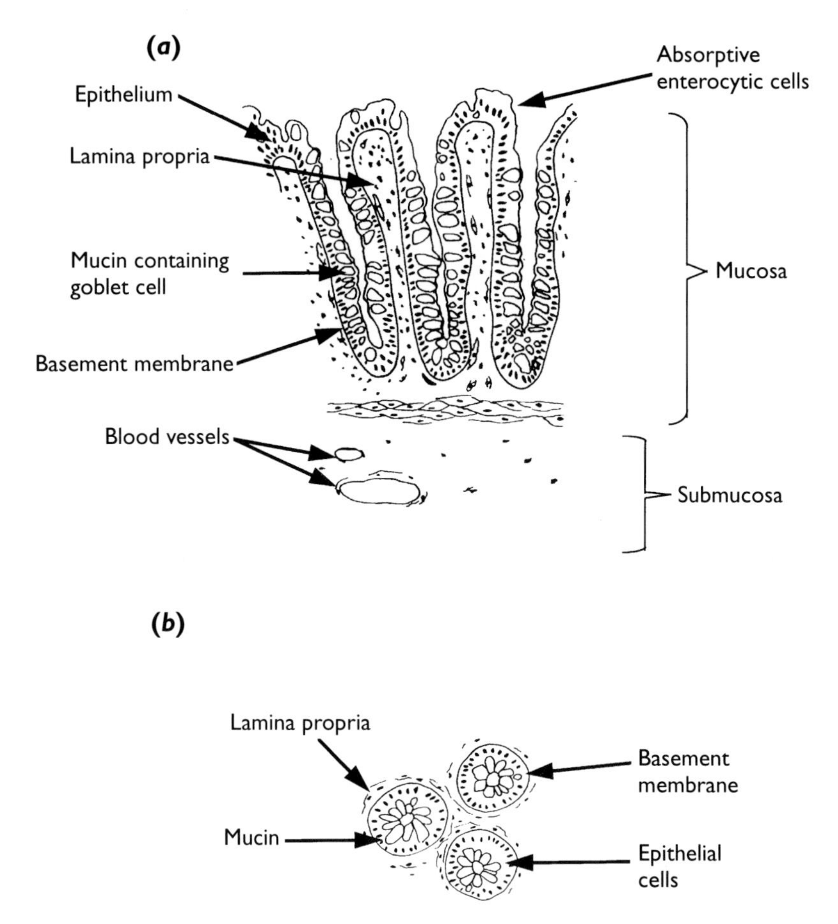

Fig. 2.3. A diagrammatic representation of the structure of normal human colonic mucosa. (*a*) Longitudinal section. (*b*) Transverse section. Note that the cell membranes are not shown; these are often indistinct in sections stained with haematoxylin and eosin. The mucosal layer is very thin (crypt length approximately 0.5 mm) and therefore care should be taken in dissecting the sample, as the stem cells are located at the base of the crypts. The lamina propria, which is located between the crypts, supports the epithelium and attaches it to the muscularis mucosae (MM). This can be a source of cell contamination, particularly pericryptal fibroblasts, which lie just below the basement membrane.

toxic to the cells) and resuspend in fresh culture medium. Transfer to a flask. For recovery from frozen, 3T3 feeder support is important. Check the culture for attachment the following day and if necessary remove any debris and replate. Check that the feeders are at sufficient density.

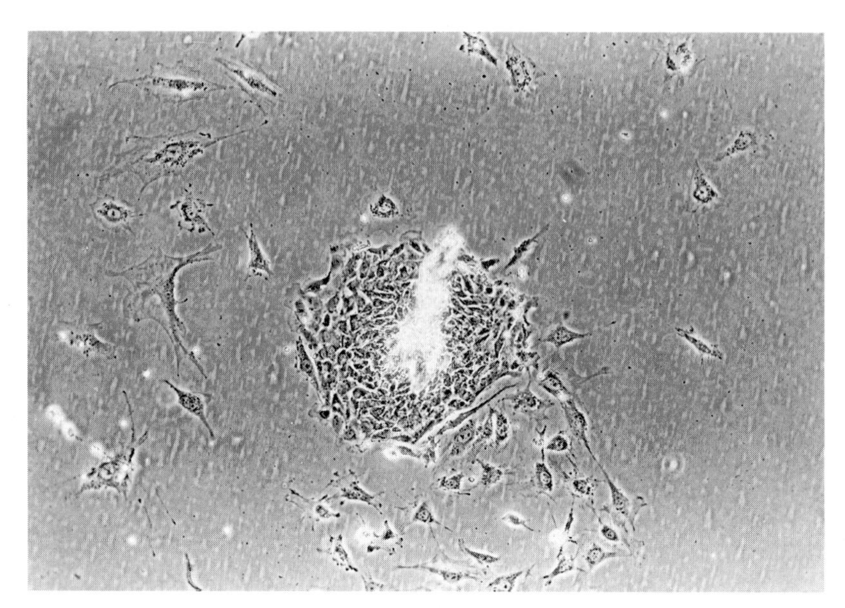

Fig. 2.4. Phase contrast photograph of a small colony of normal human colonic epithelial cells 24 h after seeding. Cells have migrated out from the central structure. Note that the 3T3 feeders would need replenishing in this culture, as they are a little sparse.

Normal colonic epithelial cell culture

Normal colonic epithelium has been successfully established from fetal tissue (Berry *et al.*, 1988). However, growth of colonic epithelium from adult normal tissue has proved difficult. Good primary cultures of normal colonic epithelial cells have been obtained in the Friedman laboratory (Buset *et al.*, 1987); however these cultures are short term, in that tissue pieces initially adhere and spread out but subsequently die after approximately 48 h in culture. The protocol we have been using is that used for culturing the adenomas and carcinomas, with minor modification.

A small sheet of intestinal mucosa is obtained from surgery. (The structure of the normal colonic mucosa is illustrated in Fig. 2.3.) One method of obtaining a mucosal sheet from a section of colon is by injecting sterile saline under the mucosa to form a 'bubble' or blister in the submucosal layer. This saline bubble does not separate the epithelium from the submucosa, but provides a firm base for removal of the mucosa. The mucosa should be removed without bursting the saline bubble. The mucosal sample is transferred into washing medium on ice for transportation to the laboratory. In preparation for culture, approximately 5–6 cm^2 can be cut at a time. The specimen is washed four times and then transferred to a sterile glass petri dish. With the mucosal side face

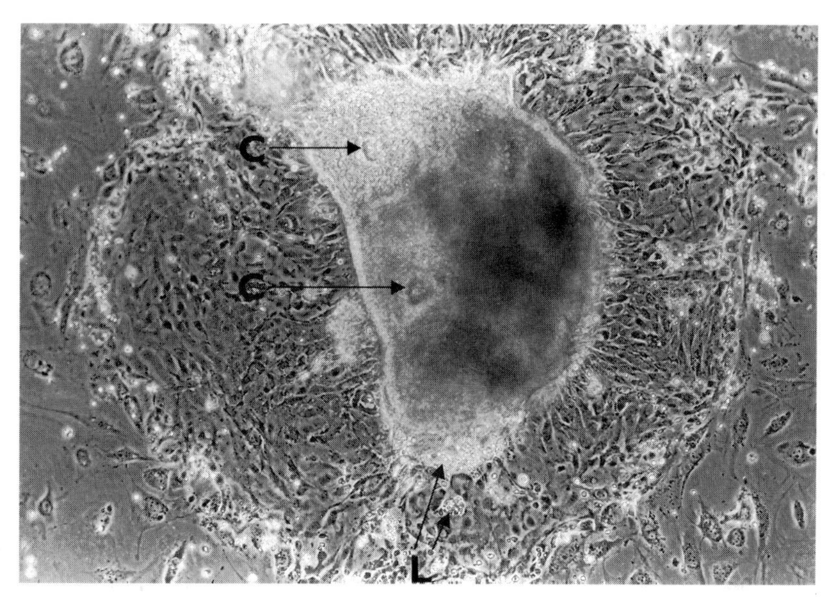

Fig. 2.5. Phase contrast photograph of a large colony of human colonic epithelial cells 24 hour after seeding (×80). Note that the colony is derived from a piece of epithelium several crypts in size and that the remnants of crypt structure (C) can still be seen in the central dense structure. The culture also contains some lymphocytes (L), which is typical of early primary cultures.

down, the submucosa, a white fibrous substance, is removed using angled scissors. Care must be taken not to cut away the base of the crypts, as it is important to retain the stem cells. This can be checked by routine histological analysis. The tissue is washed periodically to remove the dissected submucosa and to prevent the tissue from drying out. As with adenoma tissue, the normal colonic mucosa is cut using crossed scalpel blades for at least 10 min to obtain pieces small enough to pipette using a 10 ml glass pipette. The tissue is washed and put into overnight digestion in collagenase/hyaluronidase. The next day, after several washes, the tissue may need further cutting. A further 10 min of cutting can give clean mucosal pieces several crypts in size. These tissue pieces, once seeded into collagen coated flasks adhere within a few hours and start to spread out. Fig. 2.4 shows a small normal colonic epithelial colony approximately 24 h after seeding and Fig. 2.5 a larger colony after the same time period. The better colonies are derived from clumps which are composed of several crypts. In Fig. 2.5 the crypt structures can still be seen.

In our experience, the colonies start to die 48 h after seeding. The normal adult colonocytes undergo apoptosis, starting from the periphery of the colony (unpublished observations). The reason for this is as yet unclear; it may be due to limiting factors for growth and/or survival either in the medium or substrate or due to the presence of toxic factors such as transforming growth factor beta, tumour necrosis factor alpha or *fas* ligand. Longer term-culture of epithelial cells can be achieved if the separation between mucosa and the underlying muscularis mucosa and submucosa is not particularly clean. Where colonies sit on a basal layer of muscle and fibroblastic cells there may be provision of certain survival factors. The use of feeder cells from these cell lineages may improve the cultures. Clearly the further development of normal colonic primary cultures is important for our understanding of the function and regulation of the intestinal epithelial cell.

References

Bedi, A., Pasricha, P.J., Akhar, A.J., Barber, J.P., Bedi, G.C., Giardiello, F.M., Zehnbauer, B.A., Hamilton, S.R. & Jones, R.J. (1995). Inhibition of apoptosis during development of colorectal cancer. *Cancer Res.*, **55**, 1811–16.

Berry, R.D., Powell, S.C. & Paraskeva, C. (1988). *In vitro* culture of human foetal colonic epithelial cells and their transformation with origin minus SV40 DNA. *Br. J. Cancer*, **57**, 287–9.

Brattain, M.G., Marks, M.E., McCombs, J., Finely, W. & Brattain, D.E. (1983). Characterization of human colon carcinoma cell lines isolated from a single primary tumour. *Br. J. Cancer*, **47**, 373–81.

Buset, M., Winawer, S. & Friedman, E. (1987). Defining conditions to promote the attachment of adult human colonic epithelial cells. *In Vitro*, **23**, 403–12.

Chang, S.-K., Dohrman, A.F., Basbaum, C.B., Ho, S.B., Tsuda, T., Toribara, N.W., Gum, J.R. & Kim, Y.S. (1994). Localization of mucin (MUC2 and MUC3) messenger RNA and peptide expression in human normal intestine and colon cancer. *Gastroenterology*, **107**, 28–36.

Friedman, E.A., Higgins, P.J., Lipkin, M., Shinya, H. & Gelb, A.M. (1981). Tissue culture of human epithelial cells from benign colonic tumours. *In Vitro*, **17**, 632–44.

Hague, A., Manning, A.M., Hanlon, K.A., Hutschtscha, L.I., Hart, D. & Paraskeva, C. (1993). Sodium butyrate induces apoptosis in human colonic tumour cell lines in a p53-independent pathway: implications for the possible role of dietary fibre in the prevention of large-bowel cancer. *Int. J. Cancer*, **55**, 498–505.

Hall, P.A., Coates, P.J., Ansari, B. & Hopwood, D. (1994). Regulation of cell number in the mammalian gastrointestinal tract: the importance of apoptosis. *J. Cell Sci.*, **107**, 3569–77.

Jass, J.R. & Roberton, A.M. (1994). Colorectal mucin histochemistry in health and disease. *Pathol. Int.*, **44**, 487–504.

Kirkland, S.C. & Bailey, I.G. (1986). Establishment and characterization of six colorectal adenocarcinoma cell lines. *Br. J. Cancer*, **53**, 779–85.

Leibovitz, A., Stinson, J.C., McComb, W.B., McCoy, C.E., Mazur, K.C. & Mabry, N.D. (1976). Classification of human colorectal adenocarcinoma cell lines. *Cancer Res.,* **36**, 3562–9.

Paraskeva, C., Buckle, B.G., Sheer, D. & Wigley, C.B. (1984). The isolation and characterization of colorectal epithelial cell lines at different stages in malignant transformation from familial polyposis coli patients. *Int. J. Cancer,* **34**, 49–56.

Paraskeva, C., Buckle, B.G. & Thorpe, P.E. (1985). Selective killing of contaminating fibroblasts in epithelial cultures derived from colorectal tumours using an anti-Thy-1 antibody–ricin conjugate. *Br. J. Cancer,* **41**, 908–12.

Paraskeva, C., Finerty, S., Mountford, R.A. & Powell, S.C. (1989). Specific cytogenetic abnormalities in two new human colorectal adenoma-derived epithelial cell lines. *Cancer Res.*, **49**, 1282–6.

Pignatelli, M. (1993) E-cadherin: a biological marker of tumour differentiation. *J. Pathol.*, **171**, 81–2.

Williams, A.C., Harper, S.J. and Paraskeva, C. (1990). Neoplastic transformation of a human colonic epithelial cell line: *in vitro* evidence for the adenoma to carcinoma sequence. *Cancer Res*, **50**, 4724–30.

3

The pancreatic duct epithelial cell

Ann Harris

Introduction

The pancreatic duct epithelial cell is involved in a number of relatively common human pathological conditions including pancreatitis, pancreatic cancer and the inherited disorder cystic fibrosis. As a result a substantial effort has gone into developing culture systems for these cells. The ducts of the exocrine pancreas have two main functions: the delivery of digestive enzymes to the duodenum and the secretion of a bicarbonate-rich fluid that neutralises the acidic stomach products as they enter the duodenum. Bicarbonate secretion is primarily triggered by the release of secretin from endocrine cells in the duodenum, though other hormonal stimuli are also important.

The pancreatic duct system as a whole comprises less than 10% of the mass of the organ and ductal epithelial cells less than 10% of the cell number and 4% of the volume of the pancreas. The duct system is complex and contains a number of epithelial cell types. The acini contain intercalated duct cells (centroacinar cells) which lead into intralobular ducts. These in turn become interlobular ducts which increase in size as they pass towards the main pancreatic duct. The latter forms the channel between the pancreas and the duodenum. The ultrastructure of the epithelial cells lining the intralobular ducts, interlobular ducts (small and large) and the main pancreatic duct has been studied extensively in adult pancreas by Kodama (1983). Apart from indirect evidence of the main functions of each of these cell types, based on the nature of intracellular organelles, little is known about the specific properties of the different ductal epithelial cells. Generally they are characterised by apical microvilli and abundant mitochondria. Further they show high levels of expression of carbonic anhydrase and Na^+,K^+-activated ATPase. Further markers of the differentiated pancreatic duct epithelial cell will be discussed later.

Pancreatic duct epithelial cell culture systems have been set up for hamster, mouse, guinea pig and bovine pancreas (Stone *et al.*, 1978; Jones *et al.*, 1981; Hirata *et al.*, 1982; Resau *et al.*, 1983; Sato *et al.*, 1983). The cultured cells have characteristic surface microvilli and secrete mucins as they do *in vivo*. However, human pancreatic duct cells have proved difficult to isolate from the adult organ and to maintain in culture. An early report of a short-term human pancreatic duct cell culture system (Jones *et al.*, 1981) has been followed by a number of similar methods. However, the fundamental problem is that even if these cells can be purified they appear to have little if any replicative potential *in vitro*. As a result we have used mid-trimester human fetal tissues to establish longer-term cultures of pancreatic duct epithelial cells. This chapter will present in detail the methods used for the culture of fetal pancreatic duct epithelial cells and briefly present other culture systems that have been established for adult pancreatic duct cells.

The human fetal pancreas

At 9 weeks gestation the pancreas is largely composed of a simple system of branched ducts surrounded by loose interstitial tissue containing largely undifferentiated cell types. Between 12 and 14 weeks, cells from the ducts invade the interstitial tissue and start forming lobular structures. Though primitive acini may be seen after 12 weeks (Laitio *et al.*, 1974), the first mature acinar cells do not appear until somewhere near the 20th week of gestation, and the acini do not develop a lumen until the 24th week (Adda *et al.*, 1984). The mid-trimester pancreas is thus less susceptible to autolysis than post-natal pancreas, since functional digestive enzymes are not yet being synthesised.

We have established a tissue culture system for epithelial cells derived from normal mid-trimester fetal pancreas (Harris & Coleman, 1987, 1988). The system as described below results in the establishment of primary cultures of epithelial cells from about 80% of pancreases. The cells can be passaged at least two or three times before reaching terminal crisis, and may remain in culture for more than 12 weeks. Cultures have been characterised by morphological, histochemical and immunocytochemical criteria.

Culture of ductal epithelial cells from human fetal pancreas

Materials

CMRL1066 medium 10× (GIBCO BRL)

Fetal calf serum (FCS) batch tested for culture of pancreatic duct cell cultures

Penicillin 100 U/ml

Streptomycin 100 μg/ml (penicillin–streptomycin, GIBCO BRL)

L-glutamine 200 mM (100×)

Insulin (Sigma) 0.2 U/ml: reconstitute vial in 20 ml of serum-free and supplement-free medium; store aliquots at −20 °C

Cholera toxin (Sigma) 10^{-10} M: reconstitute vial in 1 ml sterile distilled H_2O; store at 4 °C

Hydrocortisone (Sigma) 1 μg/ml: make up 10 mg in 10 ml distilled H_2O, filter sterilise and store in aliquots at −20 °C

For 100 ml of medium dispense into 87 ml of sterile distilled H_2O: 10 ml CMRL1066 10×; 3 ml 7.5% sodium bicarbonate; 1 ml penicillin/streptomycin; 1 ml L-glutamine; 100 μl hydrocortisone; 160 μl insulin; 10 μl cholera toxin; 20 ml FCS (hence the final amount of FCS is less than 20%)

Primaria flasks (Falcon, Becton-Dickinson)

Collagen type IV or type VI from human placenta (Sigma)

Make up at 1 mg/ml in 1:1000 glacial acetic acid/water; place sufficient volume to just cover culture substrate and dry at 37 °C for 16 h

Collagenase type 1A (Sigma): use at 0.5 mg/ml in PBS; filter sterilise, store aliquots at −20 °C; use at 37 °C

Dispase (Boehringer Mannheim): use at 2 U/ml; make up 10 mg dispase in 50 ml of complete media; filter sterilise and use immediately at 37 °C

Fetal material

Obtain pancreas within 48 h from mid-trimester prostaglandin-induced terminations or spontaneous abortions (with local ethics committee permission).

Protocol

Remove the pancreas, still attached to the duodenum, from the fetus and wash once in tissue culture medium containing antibiotics. Further dissection is then carried out under a binocular microscope using the protection of a filtered air system. Trim away the mesentery surrounding the pancreas and then inflate the organ by injecting 1–2 ml of collagenase (Sigma type IA at 0.5 mg/ml). When visible the main duct may be microdissected from the inflated pancreas. In early mid-trimester pancreases where the main duct will not be clearly distinguished, isolate the central core of the head of the pancreas instead.

Wash microdissected tissue in tissue culture medium, cut into pieces about

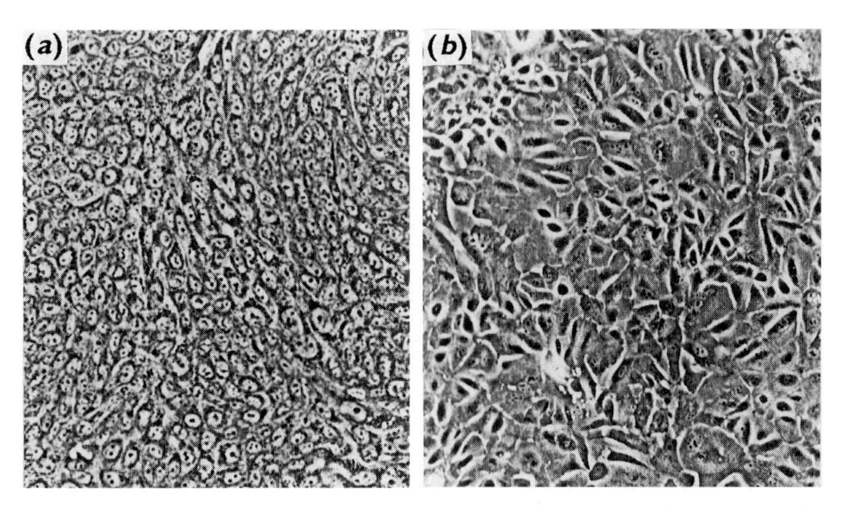

Fig. 3.1. Fetal pancreatic epithelial cell types in culture. (*a*) Tightly packed colonies of small epithelial cells (×170). (*b*) Loosely packed larger epithelial cells (×170).

1–2 mm in diameter and plate out on Primaria plastic flasks in CMRL1066 medium containing 20% FCS (GIBCO UK); penicillin (100 U/ml); streptomycin (100 μg/ml); L-glutamine (4 mM); insulin (0.2 U/ml); cholera toxin (10^{-10} M) and hydrocortisone (1 μg/ml), all from Sigma. Initially use a minimal amount of medium to just cover the explants to encourage adherence to the substrate. Once cells start migrating out of the explants add standard volumes of medium to the flask (i.e. 5 ml for a T25 flask). Maintain cultures routinely at 37 °C in a wet 5% CO_2 incubator. Various cell types will be seen migrating from the primary explants after 3–10 days.

Cell types

Two main epithelial cell types grow out of primary pancreatic explants, as illustrated in fig. 3.1: small tightly packed cells (Fig. 3.1*a*) and larger, more loosely packed cells (Fig. 3.1*b*). These two cell types are usually found together in one culture, and when this is observed the larger cells often occur at the periphery of colonies of tightly packed cells. Occasionally only one cell type may be present in a culture. Generally the tightly packed cells are the more abundant type. The cultures usually form characteristic structures *in vitro* (Fig. 3.2*a* and *b*), with areas of the smaller cells separated by large streaming cells which, though they often look fibroblastic, in fact express epithelial cell cytokeratin markers such as those detected by the monoclonal antibody LE61. (These markers will be discussed further below.)

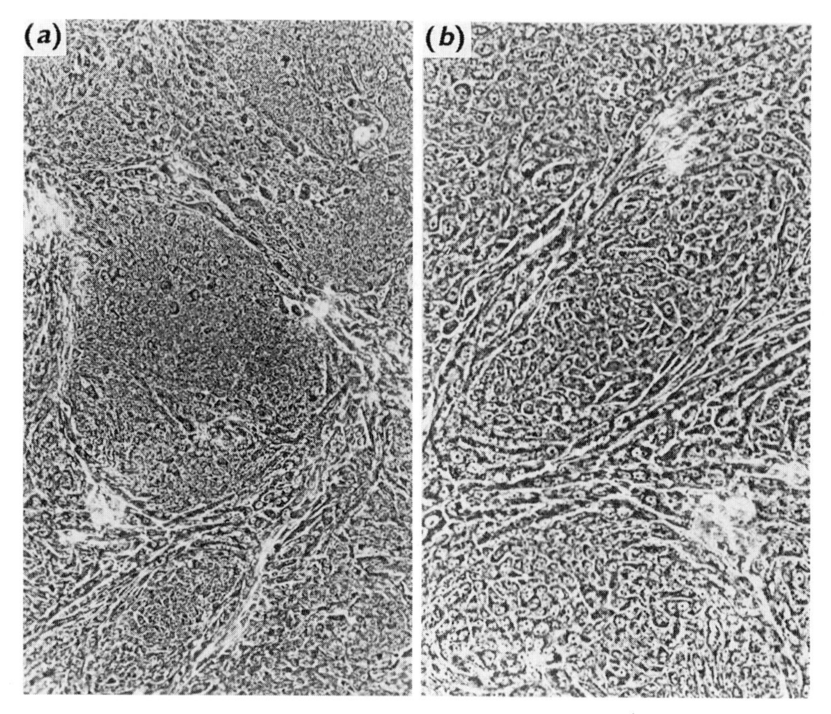

Fig. 3.2. Fetal pancreatic epithelial cell types in culture. Characteristic structures formed in primary and subsequent passages of pancreatic epithelial cells (*a* ×100, *b* ×200).

Culture media and substrates

The growth characteristics of the pancreatic epithelial cell colonies have been monitored in a variety of tissue culture media and on several different culture substrates (Harris & Coleman, 1987). CMRL1066 with 20% FCS, insulin (0.2 U/ml), cholera toxin (10^{-10} M) and hydrocortisone (1 μg/ml) is the preferred culture medium. Primaria tissue culture flasks (Falcon, Becton-Dickinson) or type IV collagen coated glass or plastic are the most efficient substrates.

Passaging of cells

Passaging of epithelial cell colonies was achieved by treating the latter with dispase, a neutral protease (Boehringer, Mannheim), at 2 U/ml in CMRL1066 medium for 15–40 min at 37 °C. The colonies were replated in 100% FCS for 24 h, then changed to 50% CMRL1066 plus 50% FCS for

24 h, and finally changed to CMRL1066 with supplements and 20% serum (the usual culture medium). Cell colonies survive passaging if they remain intact (or physically divided into smaller colonies), but not if they are disrupted into clumps of small numbers of cells. The tightly packed smaller cell type appears to be selected for by this method of passaging. Cells survive at least two or three successive passages before reaching terminal crisis, and they appear to retain constant morphological features until this time. Cultures can generally be maintained for up to 12 weeks *in vitro*.

Cryopreservation

Despite exhaustive attempts to freeze in liquid nitrogen and subsequently recover the pancreatic duct epithelial cells, we have been unable to achieve cryopreservation.

Problems encountered

Source of fetal material

One of the major factors affecting success rates for the establishment of pancreatic duct epithelial cell cultures is the source of fetal material. It appears that the two key variables are obstetric practice (methods used to induce termination) and the length of time that elapses between the start of the termination procedure and delivery of the abortus. If KCl is administered in addition to the prostaglandins used to induce delivery, this destroys the viability of the pancreatic duct epithelial cells. In addition, on occasion we see a crystalline material building up in culture media surrounding primary explant cultures. The nature of this material has not been established but viable cells never appear in these cultures. Cell cultures may be established up to 48 h after delivery of the abortus if materials are stored at 4 °C.

Age of fetal material

The level of differentiation of the fetal tissues appears to have a profound effect on the success rate in establishing cultures of ductal epithelial cells. Success rates are low before 15 weeks, primarily due to the rampant outgrowth of fibroblasts from tissue explants prior to this age. Generally, the most reliable developmental stages for the establishment of these cultures are between 18 and 21 weeks.

Contamination of primary cultures

Only rarely is contamination of cultures a problem. Usually the infection is of yeast or fungal origin and is likely to be airborne contamination entering the system during removal of the pancreas from the fetus. If this is a persistent problem it can be overcome by collecting tissue next to a bunsen flame or under the protection of a filtered air system.

Removal of fibroblasts from epithelial cell cultures

Fibroblast contamination of epithelial cell cultures frequently occurs in both primary cultures and subsequent passages. Three alternative methods have been used to eliminate fibroblasts: first, an EDTA (0.01%) wash which removes fibroblasts from the substrate before epithelial cells, so long as the former have not infiltrated the epithelial colonies; second, ricin-conjugated anti-Thy-1 (Paraskeva *et al.*,1985) (see Chapter 2) because Thy-1 antigen is expressed on fibroblasts but not on epithelial cells; third, physical removal of fibroblasts with a cell scraper. The most successful method of eliminating fibroblasts is by physical removal with a cell scraper.

Culture of ductal epithelial cells from postnatal pancreas

Materials

Eurocollins solution (Fresenius, Hamburg) or Wisconsin solution
Ficoll (Sigma)
Earle's Balanced Salt Solution (EBSS; Sigma)
EDTA (Sigma)
CMRL1066 medium (Sigma)
Hydrocortisone (Sigma): stock at 50 μg/ml; dissolve 1 ml hydrocortisone powder in 1 ml 100% ethanol, add to 19 ml 1×CMRL1066 medium; freeze in aliquots at −20 °C
Epidermal growth factor (Collaborative Biomedical): stock at 100 μg/ml in CMRL1066; freeze in aliquots at −20 °C
Glutamine (GIBCO BRL)
Soybean trypsin inhibitor (Sigma)
Insulin, transferrin and selenium (ITS; Collaborative Biomedical): 1 vial (5 l) made up in 5 ml sterile MilliQ water stored in aliquots at −20 °C
Penicillin/streptomycin (Life Technologies)
Bovine pituitary extract (BPE; Collaborative Biomedical): stock at 5 mg/ml in CMRL1066; freeze in aliquots at −20 °C

Cholera toxin (Sigma): stock at 0.5 mg/ml; 1 mg in 2 ml sterile MilliQ water; store at 4 °C
Heat-inactivated fetal bovine serum (FBS)
Aprotinin (Sigma)

Enzymes for digestion of human tissue
Collagenase: Worthington collagenase type 4 or Sigma type V (500–1000 U/ml) for whole tissue samples; Sigma type X for organ infusion
Hyaluronidase (Sigma) 300 NFU/ml, DNaseI (Sigma) 100 KU/ml: dissolve appropriate amounts in 50 ml CMRL1066 and filter sterilise; this will be sufficient for the digestion of 5 *g* tissue
Collagen (Vitrogen; Collagen Biomedical Corp.) 2 mg/ml in neutral medium salts
Hank's Balanced Salt Solution (HBSS; Sigma)

Human wash medium (HWM)
CMRL1066 supplemented with 0.2 TIU/ml aprotinin, 100 KU/ml DNaseI and 10% fetal bovine serum (FBS)

Culture medium (HM)
CMRL1066 supplemented with 1 μg/ml hydrocortisone, 10 ng/ml epidermal growth factor, 4 mM glutamine, 100 μg/ml soybean trypsin inhibitor, 5 μg/ml insulin, 5 μg/ml transferrin, 5 ng/ml selenium, 100 U/ml penicillin, 100 μg/ml streptomycin, 50 μg/ml bovine pituitary extract, 50 ng/ml cholera toxin and 5% heat-inactivated FBS

The major source of post-natal pancreatic tissue that is suitable for establishing cell cultures is a by-product of organ donor programmes. Many transplant centres, particularly in the United States and Canada, run active pancreatic islet transplant programmes. The islet purification procedure itself (Ricordi *et al.*, 1988) can generate ductal tissue isolates, alternatively small amounts of whole pancreatic tissue may be available.

Digestion of tissue and routine culture methods

Pancreatic duct cells as a by-product of the islet cell transplant procedure

Pancreas tissue should stored in isotonic maintenance fluid (Eurocollins or Wisconsin solution) until the pancreas is used. Cannulate the pancreatic duct and inject Hank's Balanced Salt Solution (HBSS) containing 2% fetal calf serum and 2 mg/ml collagenase. The amount of collagenase injected is approximately 2× the weight of the pancreas. The organ is then subjected to extensive enzymatic digestion and mechanical disruption at 37 °C (see Ricordi *et al.*, 1988). Tissue fragments are collected continuously and once free islets are seen the collagenase digestion solution in the collection chamber is gradually diluted with fresh HBSS and cooled to 4 °C. The isolation procedure is terminated 30–60 min later when islets are no longer seen in the collected sample.

Pellet the collected tissue fragments by centrifugation and load 3 ml aliquots into a 250 ml syringe with the end sealed as follows: 100 ml sterile Ficoll (lyophilised and dialysed) with a density of 1.074 g/cm^3 is pumped into the syringe. Each tissue clump aliquot is resuspended in 10 ml HBSS added to 40 ml of a Ficoll gradient (density 1.058 g/cm^3) and layered over the 1.074 g/cm^3 gradient. Spin the syringe gradients at 800 *g* for 16 min at 4 °C. The top layer (20–30 ml) containing cells, fragments and lipid droplets is discarded. The next 50–80 ml contains the islets. Islets can be compressed into a more discrete band on the gradient by using a five-step gradient between 1.033 g/cm^3 and 1.087 g/cm^3. The ductal epithelial cells do not form a tight band but are to be found in the gradient between the band of islets and the pellet of cells and tissue fragments at the bottom of the gradient. These are then centrifuged at 950 *g*, resuspended in CMRL1066, and cultured as described for the fetal cells.

Pancreatic duct cells from direct tissue digestion

Ductal epithelial cells can also be cultured directly from whole explants of pancreatic tissue. Place 1–5 *g* of pancreas tissue in a sufficient amount of Earle's Balanced Salt Solution (EBSS) with 1 mM EDTA to keep the tissue wet, and mince finely into 1–2 mm diameter fragments using sharp sterile scissors or a pair of scalpels. Aliquot the minced tissue (3 ml maximum) into 50 ml conical sterile centrifuge tubes with 10 volumes of EBSS+EDTA and agitate it on a wrist-action shaker for 10 min at 37 °C. Then pellet the mince at 500 *g* for 30 s and wash twice in human wash medium (HWM): CMRL1066 supplemented with 0.2 TIU/ml aprotinin and 100 KU/ml DNaseI.

Resuspend each pellet in 10 volumes of CMRL1066 supplemented with 500–1000 U/ml collagenase, 300 NFU/ml hyaluronidase and 100 KU/ml DNaseI. Return the mixture to a wrist-action shaker for 10 min of vigorous shaking (about 275 shakes/min). Allow the solid material to settle for 1 min then pour off the enzyme solution and recover the cells released from the tissue by centrifugation at 500 *g* for 30 s. Store the cell pellet on ice and return the enzyme solution to the minced tissue for an additional three cycles of 5 min of shaking (about 200 shakes/min) and cell recovery.

After cells have been recovered from the last cycle, discard the enzyme mix and resuspend the pellet of pancreatic tissue remaining in 10 volumes of fresh enzyme solution in a siliconised, sterile 20–30 ml glass vial with a stir bar. Continue digestion of the tissue at 37 °C in a water bath, placed on a magnetic stirrer, with an immersion heating pump to maintain the bath temperature, for 20–30 min until most of the tissue breaks down. Allow the remaining tissue to settle for 1 min and again recover the cells released from the tissue by centrifugation (as above). Discard the remaining tissue, pool all the cell samples recovered and wash them twice in EBSS +1 mM EDTA followed by a wash in HWM (to inhibit further collagenase digestion).

Resuspend the cell pellet in 10 volumes of HWM and pass it through a 140–190 μm stainless steel tissue sieve. Wash the tissue on the sieve with HWM then recover the cell clumps by back-washing them off the mesh, into a sterile tube or beaker, with HWM dispensed from a 10/20 ml syringe and a 21 gauge needle. Pellet the cell clumps by centrifugation (500 *g* for 30 s) and resuspend them in liquid collagen gel (2.2 mg/ml collagen in neutral medium salts). Dispense 2 ml of the clump suspension onto an equal volume of gelled collagen in a 60 mm tissue culture dish, allowing 100 μl of the initial cell clump volume per dish. Place dishes in a 37 °C incubator until the collagen solidifies and then overlay them with 5 ml HM. Change the medium after 24 h and then every 48 h for the subsequent 7–10 days, during which time cystic forms appear in the cell clumps trapped in the collagen gel. The main advantages of growing cystic forms prior to generating monolayer cultures are as follows: (1) the ducts have time to recover while surrounded by the collagen support; (2) acini and islets that may be contaminating initial cultures die during this procedure and fibroblast growth is not encouraged; (3) viable ducts can readily be harvested from the collagen gels.

To initiate monolayer cultures, dissolve the collagen gel with an equal volume of 0.1% collagenase in CMRL1066 (this usually takes about 30 min at 37 °C). Collect the cystic clumps in a sterile, siliconised pasteur pipette with the aid of a dissecting microcope. Wash the cysts twice in HWM with collection by centrifugation at 500 *g* for 30 s and resuspend them in HM distributed onto thin collagen gels (on tissue culture flasks or permeable filters) or on Primaria plastic flasks or dishes. Monolayer cultures then develop (Fig. 3.3*b* and *c*) . Generally 5 *g* of pancreas tissue yields up to 8×60 mm diameter dishes of pancreatic duct cells at confluence. It has not proved possible to passage these cells.

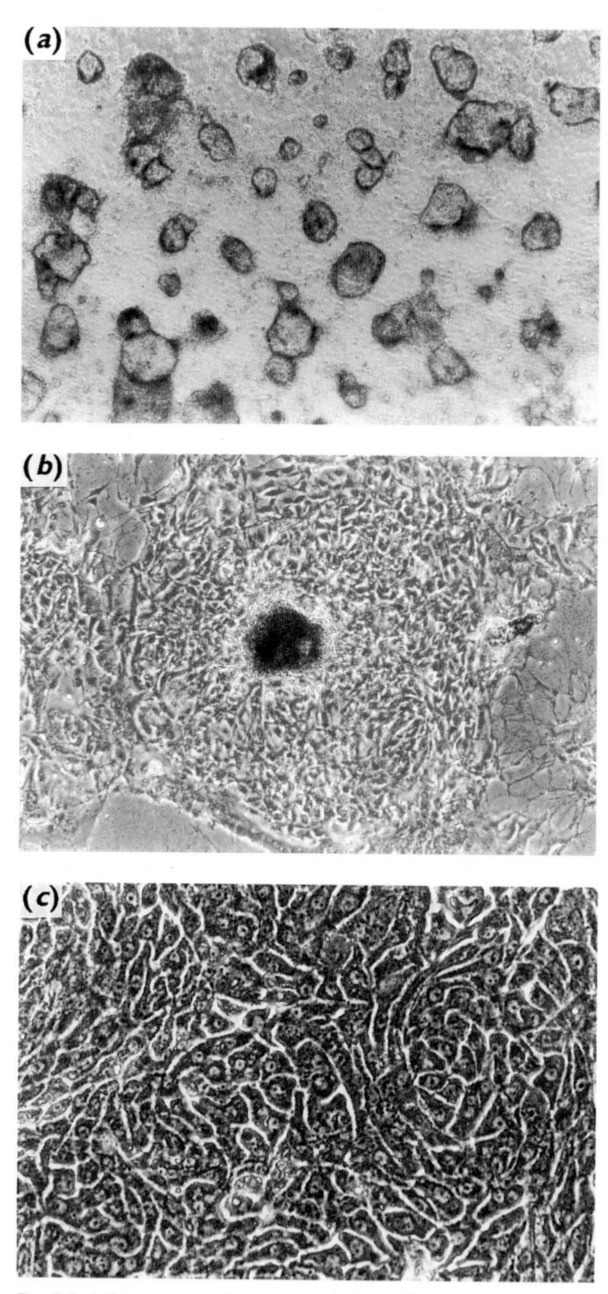

Fig. 3.3. Adult pancreatic duct epithelial cell cultures (kindly provided by C. Kolar and T.A. Lawson). (*a*) Cystic forms in a collagen gel (×20). (*b*) Early monolayer culture: lobed cyst with monolayer outgrowth of epithelial cells (×30). (*c*) Monolayer culture (×150).

Problems encountered

Tissue viability

Some tissue samples show autolysis early in the process, which is probably dependent on cold ischaemia time and is more common in tissue from older donors. This autolysis cannot be interrupted but the yield of viable cells may be improved by removing part of the enzyme mixture and renewing it with fresh solutions periodically during the period of wrist-action shaking.

Collagenase

There is substantial variation in the efficacy of different batches of collagenase. The range of units used in the tissue digestion (500–1000 U/ml) is a function of this. Titrate each new batch of collagenase starting with 500 U/ml and increase until adequate digestion is achieved in the time scale described above.

Age of donors

Pancreatic tissue from donors over the age of 35 years shows a marked reduction in the viability and proliferation of the ductal cells. With donors above the age of 45 years one is unlikely to obtain viable cells.

Markers for the characterisation of the cultured cells

An increasing number of markers for differentiated pancreatic duct epithelial cells have been reported (Hootman & de Ondarza, 1993; Githens, 1994). Data on the use of only a few of these will be presented below as they give adequate characterisation of the ductal epithelial cell nature of the cultured cells.

Immunocytochemistry

Cytokeratins

Both small and large epithelial cell types express the mixture of cytokeratin intermediate filament proteins characteristic of simple epithelia. These are detected by monoclonal antibodies specific for cytokeratins 8 and 18 (commercially available). In addition the fetal pancreatic duct cells express cytokeratin 19, for which specific monoclonal antibodies are also commercially available. In adult pancreas, exocrine acini express predominantly cytokeratins 8 and 18, while microdissected pancreatic ducts also contain cytokeratins 7 and 19 (Moll *et al.*, 1983).

Carbonic anhydrase

The only parts of the adult pancreas that have been shown in tissue sections to express carbonic anhydrase are the pancreatic ducts (Kumpulainen & Jalovaara, 1981; Spicer *et al.*, 1982). Fetal pancreatic duct cells in culture also express carbonic anhydrase.

Mucins

The MUC1 mucin gene product is expressed in cultured fetal pancreatic duct cells. There exist a variety of monoclonal antibodies that are specific for the MUC1 mucin, including monoclonal antibodies Ca2 (Bramwell *et al.*, 1983), HMFG2 (Burchell *et al.*, 1983) and DU-PAN-2 (Borowitz *et al.*, 1984).

Reverse transcription–polymerase chain reaction (PCR)

Small amounts of RNA can be prepared from the cultured epithelial cells to ensure that they are transcribing some of the genes involved in the differentiated functioning of pancreatic duct epithelial cells. One of these genes is the cystic fibrosis transmembrane conductance regulator (CFTR) gene that is expressed at a high level in this part of the pancreas. The following is a brief summary of the assay that we use for the detection of CFTR mRNA (Chambers & Harris, 1993).

Reverse transcription (RT)–PCR

cDNA synthesis. Resuspend 1 μg total RNA in 5 μl diethyl pyrocarbonate (dEPC)-treated water for each reaction. Add 4.5 μl 'premix 1', containing 0.5 μl of a 3′-oligonucleotide primer specific for CFTR (Chambers & Harris, 1993) (100 ng/μl) and 4 μl T0.1E (10 mM Tris, 0.1 mM EDTA), to the RNA, cover with mineral oil and incubate at 65 °C for 10 min. Transfer the sample to ice and add 10.5 μl 'premix 2', containing 4 μl of 5×RT buffer (250 mM TRIS.HCl pH 8.5, 375 mM KCl, 50 mM dithiothreitol (DTT), 15 mM $MgCl_2$), 5 μl of a solution of 5 mM dNTPs, 0.5 μl RNAse inhibitor (human placental, Boehringer), 0.5 μl dH_2O and 1 μl reverse transcriptase (MMLV 200 U/μl, GIBCO), below the mineral oil and incubate the mixture at 42 °C for 1 h. Controls in which either no RNA or no reverse transcriptase is added must be included to indicate the presence of contamination.

Amplification. Add the entire cDNA to be amplified (20 μl) to a PCR premix containing 5 μl of the 5′ primer and 4.5 μl of the 3′-oligonucleotide primer (100 ng/μl), 5 μl 10× PCR buffer (670 mM TRIS.HCl pH 8.8, $(NH_4)_2SO_4$ 166 mM, $MgCl_2$ 67 mM), 5 μl of a solution of 5 mM each dNTP, 1.7 μl 0.5% bovine serum albumin, 0.7 μl 5% mercaptoethanol and 0.6 μl of *Taq* DNA polymerase, made up to 50 μl with T0.1E. The premix should be exposed to ultraviolet light for 5 min before addition of the *Taq* DNA polymerase. Perform 30 cycles of PCR. For the standard RT–PCR reaction that we use for the detection of CFTR message, primers EIR (5′AGATTCTC-CAAAGATATAGC3′) and E1L (5′GAAATGTTGTCTAATATGGC3′) are used, with an annealing temperature of 60 °C and 5 m extension at 72 °C. We generally include a set of primers for a housekeeping gene in order to verify that the RNA was intact and all other components in the RT–PCR reaction are functioning efficiently.

Concluding remarks

A tissue culture system for epithelial cells derived from normal mid-trimester fetal pancreases is described here. These cells could be either ductal epithelial cells or acinar cells, which share common developmental pathways and have many structural and functional characteristics in common. On the basis of morphological, histochemical and immunocytochemical characteristics it is likely that these cells are ductal in origin. It is possible that they do not correspond directly to one of the differentiated duct cell types seen in adult human pancreas, but rather are developmental precursors of these cells. However, the epithelial cells reported here do show many characteristics of differentiated interlobular duct cells (Kodama, 1983). Cultures derived from adult pancreatic ducts show very similar differentiated characteristics. Though substantially more cells can be obtained from an adult pancreas, these cells do not replicate efficiently in culture and so cannot be expanded, nor can they be passaged.

Acknowledgements

This cell culture system was established through grants from the Cystic Fibrosis Research Trust, UK. The author particularly wishes to thank Mrs Lindsay Collins, Drs Mary Seller, Sylvain Phaneuf and Judith Asselin for their help; also Ms Carol Kolar and Dr Terry Lawson, Eppley Cancer Institute, Omaha, NE, USA, for allowing the inclusion of their unpublished methods for culturing adult pancreatic duct cells.

References

Adda, G., Hannoun, L. & Loygue, J. (1984). Development of the human pancreas: variations and pathology. A tentative classification. *Anat. Clin.*, **5**, 275–83.

Borowitz, M.J, Tuck, F.L., Sindelar, W.F., Fernsten, P.D. & Metzgar, R.S. (1984). Monoclonal antibodies against human pancreatic adenocarcinoma: distribution of DU-PAN-2 antigen on glandular epithelia and adenocarcinomas. *J. Natl. Cancer Inst.*, **72**, 999–1003.

Bramwell, M.E., Bhavanandan, V.P., Wiseman G. & Harris H. (1983). Structure and function of the Ca antigen. *Br. J. Cancer*, **48**, 177.

Burchell, J., Durbin, H. & Taylor-Papadimitriou, J. (1983). Complexity of expression of antigenic determinants recognised by monoclonal antibodies HMFG1 and HMFG2, in normal and malignant human mammary epithelial cells. *J. Immunol.*, **131**, 508–13.

Chambers, J.A. & Harris, A. (1993). Expression of the cystic fibrosis gene and the major pancreatic mucin gene (MUC1) in primary human ductal epithelial cells and in pancreatic duct cell lines. *J. Cell Sci.*, **105**, 417–22.

Githens, S. (1994). Pancreatic duct cell cultures. *Annu. Rev. Physiol.*, **56**, 419–43.

Harris, A. & Coleman, L. (1987). Establishment of a tissue culture system for epithelial cells derived from human pancreas: a model for the study of cystic fibrosis. *J. Cell Sci.*, **87**, 695–703.

Harris, A. & Coleman, L. (1988). Cultured epithelial cells derived from human fetal pancreas as a model for the study of cystic fibrosis: further analyses on the origins and nature of the cell types. *J. Cell Sci.*, **90**, 73–7.

Hirata, K., Oku, T. & Freeman, A.E. (1982). Duct, exocrine, and endocrine components of cultured fetal mouse pancreas. *In Vitro*, **18**, 789–99.

Hootman S.R. & de Ondarza, J. (1993). Overview of pancreatic duct physiology and pathophysiology. *Digestion*, **54**, 323–30.

Jones, R.T., Hudson, E.A. & Resau, J.H. (1981). A review of *in vitro* and *in vivo* culture techniques for the study of pancreatic carcinogenesis. *Cancer*, **47**, 1490–6.

Kodama, T. (1983). A light and electron microscope study of the pancreatic ductal system. *Acta Pathol. Jpn.* **33**, 297–321.

Kumpulainen, T. & Jalovaara, P. (1981). Immunohistochemical localization of carbonic anhydrase isoenzymes in the human pancreas. *Gastroenterology*, **80**, 796–9.

Laitio, M., Lev, R. & Orlic, D. (1974). The developing human fetal pancreas: an ultrastructural and histochemical study with special reference to exocrine cells. *J. Anat.*, **117**, 619–34.

Lane, E. (1982). Monoclonal antibodies provide specific intramolecular markers for the study of epithelial tonofilament organization. *J. Cell Biol.*, **92**, 665–73.

Moll, R., Krepler, R. & Franke, W.W. (1983). Complex cytokeratin polypeptide patterns observed in certain human carcinomas. *Differentiation*, **23**, 256–69.

Paraskeva, C., Buckle, B.G. & Thorpe, P.E. (1985). Selective killing of contaminating human fibroblasts in epithelial cultures derived from colorectal tumours using an anti thy-1 antibody–ricin conjugate. *Br. J. Cancer*, **51**, 131–4.

Resau, J.H., Hudson, E.A. & Jones, R.T. (1983). Organ explant culture of adult Syrian golden hamster pancreas. *In Vitro*, **19**, 315–25.

Ricordi, C., Lacey, P.E., Finke, E.H., Olack, B.J. & Scharp, D.W. (1988). Automated method for isolation of human pancreatic islets. *Diabetes*, **37**, 413–20.

Sato, T., Mamoru, S., Hudson, E.A. & Jones, T. (1983). Characterization of bovine pancreatic ductal cells isolated by a perfusion–digestion technique. *In Vitro*, **19**, 651–60.

Spicer, S.S., Sens, M.A. & Tashian, R.E. (1982). Immunocytochemical demonstration of carbonic anhydrase in human epithelial cells. *J. Histochem. Cytochem.*, **30**, 864–73.

Stone, G.D., Harris, C.C., Bostwick, D.G., Jones, R.T., Trump, B.F., Kingsbury, E.W., Fineman, E. & Newkirk, C. (1978). Isolation and characterization of epithelial cells from bovine pancreatic duct. *In Vitro*, **14**, 581–90.

4

Renal and bladder epithelial cells

Stanley J. White, John R.W. Masters and Adrian S. Woolf

Introduction

The diverse functions of kidney epithelia include the sieving of plasma by glomeruli to produce a near protein-free ultrafiltrate, and the modification of urine by tubules which reabsorb ions and water and also secrete ions. During renal development nephrons differentiate from mesenchymal cells, apart from the collecting ducts which are branches of the ureteric bud. Although these precursors can be cultured (Woolf *et al.*, 1995), this chapter will focus on the isolation and propagation of epithelial cells from the adult kidney. The epithelia of the renal pelvis, ureter and bladder are fairly homogeneous but differ structurally from cells which line the nephron tubules, reflecting the primary function of the urothelium as a water-tight conduit for urine.

Research applications of culture of renal and bladder epithelia

The culture of isolated kidney and bladder epithelia has widespread application to both basic and applied science. Cell culture allows us to study cells in isolation from confounding influences which occur *in vivo*, such as blood and urine flow and the proximity of other cell types such as fibroblasts and endothelial cells.

Growth, differentiation and tumour formation

Mitoses are only rarely observed in the normal adult kidney and urinary bladder, and the acquisition of increased proliferation has relevance to the pathogenesis of tumours, regeneration after toxicity, and also the polycystic kidney diseases. In the latter disorders renal epithelial cells appear to have escaped the normal curbs to proliferation. Using tissue culture it is possible to study the effects of defined growth factors on cell differentiation and proliferation. The genesis of polarity can be examined by studying apical and basolateral cell functions when growing epithelial monolayers on semipermeable supports. In addition, renal tubules and cysts can be reconstituted *in vitro* (Humes & Cieslinski, 1992; Woo *et al.*, 1994). As an example of study of tumour formation, Reznikoff *et al.* (1993) immortalised human urothelial cells to derive a line which became cytogenetically unstable after serial passaging, and transformation was associated with chromosome losses.

The physiology of water and ion transport

Aberrant physiology is central to tubular diseases, such as Fanconi syndrome (Racusen *et al.*, 1991), when the proximal tubule fails to reabsorb phosphate and bicarbonate, and nephrogenic diabetes insipidus, when the collecting ducts are unable to concentrate urine. Vectorial transport can be studied in monolayers using Ussing-type chambers and also in epithelial cysts reconstructed *in vitro*. The measurement of intracellular concentrations of ions and also pH has added a new dimension to the study of renal physiology, and ion channels can be studied in single cells by the patch-clamp technique.

Cytotoxicity

Kidney tubules can be damaged by ischaemic and pharmacological insults which are major causes of acute kidney failure. These events can be mimicked *in vitro* using cultured tubule epithelia and agents can be designed to minimise renal toxicity. Similarly, glomerular epithelial cells can be damaged by drugs, such as puromycin, causing the loss of selectivity of the glomerular barrier leading to leakage of protein into the filtrate; again, these events can be studied in tissue culture.

Cell transplantation

A number of groups are attempting to grow urothelial cells for autologous transplantation for hypospadias repair and bladder augmentation. It is now possible to grow urothelial cells *in vitro* to produce enough tissue for a graft, but there is still much progress that needs to take place concerning substrates and surgical techniques before such grafts become a routine part of clinical practice.

Culture of renal epithelial cells

General considerations on characterisation

When characterising a renal epithelial line in culture, a battery of tests should be performed because there are few, if any, definitive markers for any one type of cell. It should also be realised that although many of the differentiated features listed below are present in the primary culture, they may eventually be lost. The mature kidney contains over ten types of epithelial cell. With the exception of glomerular podocytes, renal epithelia share many features in common. They are polarised, with the abluminal (basal) plasma membrane attached to a basement membrane and the luminal (apical) membrane supporting microvilli or cilia. Tight junctions are located at the junction of the apical and lateral membranes and limit flow through the paracellular pathway. Below these zonula occludens are located the zonula adherens, the desmosomes and the gap junctions. Renal epithelia have a highly organised cytoskeleton with actin filaments around the perimeter of the cytoplasm and cytokeratin intermediate filaments in the cell body. For general details of renal epithelial cell characterisation see Kreisberg & Wilson (1988), Wilson (1991) and McAteer *et al.* (1991), in addition to specific references cited below.

Glomerular visceral epithelial cells

The glomerular visceral epithelial cells are called podocytes because their cell bodies extend 'foot processes' which stand on the surface of endothelial cells. Podocytes and endothelial cells form the barrier through which plasma passes to form an ultrafiltrate which enters Bowman's space. Adjacent foot processes are joined by a membrane called the slit diaphragm. Therefore, the morphology of the podocyte differs from that of other renal epithelia. Moreover, these cells express vimentin, intermediate filaments characteristic of mesenchymal rather than epithelial cells, and WT1, a gene coding for a transcription factor in the embryonic kidney. *In vivo*, podocytes also express gp330, the Heyman Nephritis antigen, podocalyxin, a sialoprotein which covers podocytes (Kerjaschki *et al.*, 1986), and they bind to wheat germ (*Triticum vulgaris*) lectin which has affinity for *N*-acetyl-β-D-glucosaminyl residues and *N*-acetyl-β-D-glucosamine oligomers. The epithelial cells lining Bowman's capsule are called the parietal cells, and they have the general structure of classical renal epithelial cells joined by desmosomes and expressing cytokeratin (Weinstein *et al.*, 1992).

Proximal tubules

Cells in the proximal tubules of the nephron reclaim the bulk of the filtrate and catabolise large molecules. These columnar cells contain many mitochondria, lysosomes and vesicles and possess a rich brush border of microvilli. Proximal tubules express gp330, alkaline phosphatase, 5′-nucleotidase, γ-glutamyl transpeptidase, leucine aminopeptidase and maltase, and they bind to the asparagus pea (*Lotus tetragonolobus*) lectin which has high affinity for α-L-fucose residues. They express various transporters and co-transporters including Na^+–glucose, Na^+–amino acid and phosphate–glycoprotein. Their Na^+-dependent phosphate transport system is inhibited by parathyroid hormone (PTH), which stimulates cyclic adenosine monophosphate (cAMP). Proximal tubule cells express a phlorizin-sensitive, Na^+-dependent uptake of α-methyl-D-glucoside, a substrate for the Na^+-dependent hexose transport system. These cells are reported to proliferate in response to fibroblast growth factor, epidermal growth factor (EGF), insulin-like growth factor and cholera toxin. *In vitro*, they endocytose lucifer yellow and horseradish peroxidase in a non-saturable, energy-dependent process. A recent study covers many of these aspects of characterisation (Romero *et al.*, 1992).

Thick ascending loop of Henle

The remainder of the glomerular filtrate enters the descending loop of Henle and becomes hypotonic as it passes up the ascending loop. Cells of the latter segment contain many mitochondria and have high levels of Na^+–K^+-ATPase located in the basolateral plasma membrane. Pre-pro EGF is synthesised by these cells and is targeted to the apical membrane, as is Tamm–Horsfall protein. Cells of the ascending loop of Henle respond to calcitonin by increasing levels of cAMP.

Collecting tubules

The functions of this distal segment include potassium secretion, calcium reabsorption and H^+ secretion and reabsorption, and it is composed of different types of cells. Principal cells have few organelles and respond to arginine vasopressin (AVP) and aldosterone. They reabsorb Na^+ using a basolateral Na^+–K^+-ATPase and apical amiloride-sensitive Na^+ channel, and they secrete K^+ using a Ba^{2+}-sensitive apical channel. In contrast, intercalated cells are rich in mitochondria and cytochrome oxidase: α cells reabsorb bicarbonate using an apical H^+-ATPase, basolateral Cl^-/HCO_3^- exchanger, and an intracellular carbonic anhydrase, while β cells secrete bicarbonate. Peanut (*Arachis hypogaea*) lectin binds to the bicarbonate-secreting intercalated cells but may also stain cells which express band 3, a marker for acid-secreting cells. In tissue culture a combination of aldosterone and insulin increases the number of peanut agglutinin-binding cells (Minuth *et al.*, 1994) and β intercalated cells have been observed to differentiate into α cells and principal cells (Fejes-Toth *et al.*, 1992). Cultures from the cortical collecting duct form a tight monolayer with a high electrical resistance and a lumen negative voltage, both characteristics of this segment *in vivo*.

Overview of tissue culture of renal epithelial cells

Isolation of renal epithelial cells

The method used for isolating renal epithelial cells depends on the specific type of renal epithelium being studied as well as the purity of the preparation and quantity of cells required. Enzymatic dissociation of renal cortex yields a predominant population of proximal tubule cells (McAteer *et al.*, 1991). Microdissection of individual tubules is time consuming and of low yield but will give a pure population of cells (Merot *et al.*, 1989; Wilson, 1991). When large numbers of rarer cells are required, fluorescence-activated cell sorting in combination with immunodissection–solid phase immunoadsorption has been used to isolate cells from the collecting duct (Fejes-Toth & Naray-Fejes-Toth, 1991) and proximal tubules. Another method, which involves the separation of segments on a density gradient, is described below (White *et al.*, 1992). Fresh human kidney tissue is relatively scarce and it is of note that a few viable proximal tubular cells are shed into the urine and these can be grown successfully in culture (Racusen *et al.*, 1991).

Obtaining pure populations of renal epithelial cells

The bulk of renal epithelia are composed of proximal tubules, while other epithelial cells, such as the intercalated cell, are much rarer. There is therefore a numerical bias against the investigator being able to isolate large numbers of some types of renal epithelial cell in primary culture. In some experiments, the presence of a few per cent of contaminating cells is of little relevance; on the other hand, experiments on single cells or those using sensitive techniques such as the polymerase chain reaction demand a pure population. Additionally, a minor population of fibroblasts and mesangial cells which may be present in the original tubule or glomerular isolate may overrun the epithelial culture, especially in the presence of serum. Fortunately, renal epithelial cells will often proliferate in serum-free media supplemented with molecules such as insulin, transferrin, prostaglandin E_1, hydrocortisone and triiodothyronine, and such 'defined' media, two of which are described below, can reduce overgrowth by fibroblasts.

Generation of large numbers of renal epithelial cells

Normal adult renal epithelial cells stop dividing after a few passages. Spontaneously immortalised renal epithelia are available (e.g. Madin–Darby canine kidney cells) but none of the available lines faithfully represents the phenotype of specific renal epithelia found *in vivo*. It is possible to introduce immortalising genes (e.g. the large T antigen of the SV 40 genome) by DNA transfection or viral transduction of primary cultures, or by direct derivation of cell lines from mice transgenic for oncogenes. While this strategy facilitates cell cloning, immortalisation may preclude a fully differentiated phenotype. Recently, it has become feasible to introduce 'conditionally immortalising' genes which, when active, allow cell proliferation; the immortalising gene can then be 'turned off' to enhance differentiation (Prie *et al.*, 1991; Woolf *et al.*, 1995). Other strategies to prevent the loss of differentiation in culture include growing cells on basement-membrane-like substrates such as laminin and collagens, and the addition of differentiating agents including retinoic acid and epidermal growth factor (Humes & Cieslinski, 1992; Nosaka *et al.*, 1993).

Protocol for culture of proximal and distal nephron cells

Overview

The following protocol allows separation of proximal and distal nephron segments from rabbit kidney and it works well for other species such as the rat and human kidney (White *et al.*, 1992). However, the growth and functional properties of such cultures may not be the same as for the rabbit, which can be grown in serum-free media (see below).

Tissue collection

Materials

Sterile 100 ml beaker

Sterile dissecting scissors and forceps

Solutions and reagents

All sterilised by filtration (0.2 μm). All reagents are obtainable from Sigma unless otherwise stated.

HEPES (*N*-2-hydroxyethylpiperazine-*N'*-2-ethanesulphonic acid) solution is composed of:

NaCl 118.0 mM
KCl 4.7 mM
$CaCl_2$ 1.3 mM
$MgSO_4$ 1.3 mM
KH_2PO_4 1.2 mM
Na glucuronate 20.0 mM
Glucose 5.0 mM
Glycine 5.0 mM
HEPES 5.0 mM
Penicillin 10^5 U/l
Streptomycin 100 mg/l
Solution is buffered with NaOH to pH 7.4 at 4 °C

Tissue preparation

Protocol

1 Remove kidney as aseptically as practicable from a freshly killed male or female New Zealand white rabbit (1–2 kg) [pentobarbitone (200 g/l) by intravenous injection] and place the organs in ice-cold HEPES solution.
2 Rinse kidney in ice-cold HEPES solution, three times. Trim excess fat, decapsulate and make thin (250 μm) cortical sections with a Stadie-Riggs microtome (Thomas Scientific). Store sections in ice-cold HEPES.

Enzymatic dispersion of kidney cortex and isolation of tubule fragments (Fig. 4.1)

Materials

Sterile 10 cm glass petri dishes
Two pairs of sterile disposable scalpels (no. 11 blades)
Trypsinising flask (Wheaton)
Water bath at 37 °C
Cell dissociation sieves (Sigma) 40, 200 and 300 mesh sterilised by autoclave
50 ml sterile polypropylene centrifuge tubes
50 ml polycarbonate centrifuge tubes for high-speed centrifuge (Beckman)

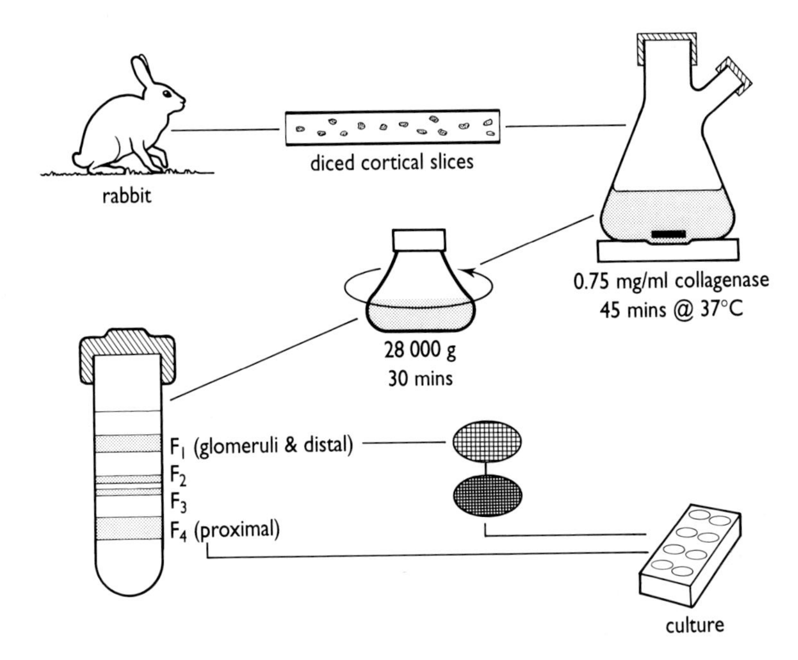

Fig. 4.1. Scheme for isolation of proximal and distal tubule fragments. Enzymatic dispersion of kidney cortex and isolation of tubule fragments.

Solutions and reagents

HEPES solution

'RKI medium': a 1:1 mix of Ham's F-12 and Dulbecco's Modified Eagle Medium (DMEM) high glucose with no HCO_3^- supplemented with:

Human transferrin 5 mg/l

Insulin 5 mg/l

Sodium selenite 50 nM

Hydrocortisone 50 nM

Penicillin 10^5 U/l

Streptomycin 100 mg/l

1 ml HEPES containing 7.5 mg collagenase (Worthington type IV)

10% bovine serum albumin (BSA) in HEPES

5% BSA in HEPES

Percoll

2× standard concentration of HEPES solution

Protocol

1 Place sections in 4.5 ml HEPES. Add 4.5 ml RKI medium and 0.25 ml 10% BSA. Disrupt sections with sterile scissors and add an aliquot of 1 ml collagenase solution. Transfer the suspension to a sterile trypsinising flask containing a stirring bar and incubate with stirring for 45 min at 37 °C. Gas with 95% O_2:5% CO_2.
2 Add 10 ml cold HEPES (reaction stopper) and strain the suspension through a sterile coarse strainer (40 mesh) into a 250 ml beaker and then transfer the filtrate into two 50 ml tubes.
3 Centrifuge the tubes at 1000 *g* for 3 min at 4 °C. Discard the supernatant and resuspend the pellet in 30 ml cold HEPES in a single 50 ml tube.
4 Repeat step 3 three times. Prior to the third spin resuspend in 5% BSA. Stand the tubes at 4 °C for 5 min. Spin at 1000 *g* for 3 min.
5 Discard the supernatant, resuspend the pellet in 80 ml Percoll mixture (40 ml Percoll and 40 ml HEPES 2× concentration). Pipette the suspension into polycarbonate tubes. Balance to 0.1 g.
6 Spin at 20 000 *g* for 30 min at 4 °C.

Cortical collecting duct (CCD) culture

Protocol

1 Carefully remove the top (F1) layer and place in a 50 ml tube. Filter through 200 mesh filter with 200 ml ice-cold HEPES into a 200 ml beaker. Pass the filtrate through the 300 mesh filter with 200 ml of ice-cold HEPES (400 ml in total). This procedure removes glomeruli and some contaminating proximal tubule segments from the layer. These glomeruli can be used for glomerular cell culture as described elsewhere (Kreisberg & Wilson, 1988).
2 Pour the filtrate, which consists of small fragments of distal nephron, into 8×50 ml tubes and spin at 1000 *g* for 1.5 min.
3 Resuspend the pellets in 30 ml RKI medium and spin at 1000 *g* for 1.5 min.
4 Resuspend the final pellet in 18 ml RKI medium and plate out in two or three 24 well dishes at a density of 5×10^4 fragments per well. Alternatively, for functional studies, seed the resulting suspension onto rat-tail collagen coated permeable 10 mm inserts (Costar Snap Well from Falcon).

Proximal tubule (PT) culture

Solutions and reagents

'Proximal tubule RKI medium' consists of a 1:1 mix of Ham's F-12 and DMEM, low glucose and no HCO_3^- supplemented with:

Human transferrin 35 mg/l
Insulin 5 mg/l
Sodium selenite 25 nM
Ethanolamine 20 μM
Penicillin 10^5 U/l
Streptomycin 100 mg/l

Protocol

1 Remove the bottom layer (F4) of the Percoll gradient which consists of 98% pure proximal tubules. The fragments appear to be a mixture of all three proximal segments (S1–S3). Wash twice with HEPES and then once with 'Proximal tubule RKI medium'. Resuspend in an appropriate volume for 5×10^4 fragments per well or 10 mm insert as described above.

Establishment and maintenance of cultures

Cells are cultured at 37 °C in an atmosphere of 95% air and 5% CO_2. CCD cultures should be fed after 48 h, at which time antibiotics are withdrawn. PT cultures are fed after 72 h. Subsequently, feeding is typically carried out every 48 h for the lifetime of the cultures. Both types of culture are usually confluent within 4–6 days. They display differentiated transepithelial function (maintenance of a transepithelial potential) for up to 2 weeks.

Characterisation

The Percoll gradient method of isolation has the advantage of yielding large quantities of material but the disadvantage of uncertainty regarding the absolute specificity of cell type. However, in our hands, the two cultures show many of the differentiated properties as described below (White *et al.*, 1992):

1 The proximal tubule monolayers form a low-resistance ($<10\ \Omega \cdot cm^2$) uniform epithelium (Fig. 4.2) and possess a well-developed brush border. They undergo apical membrane depolarisation on addition of glucose (indicative of Na^+-coupled co-transport). Cyclic AMP synthesis is

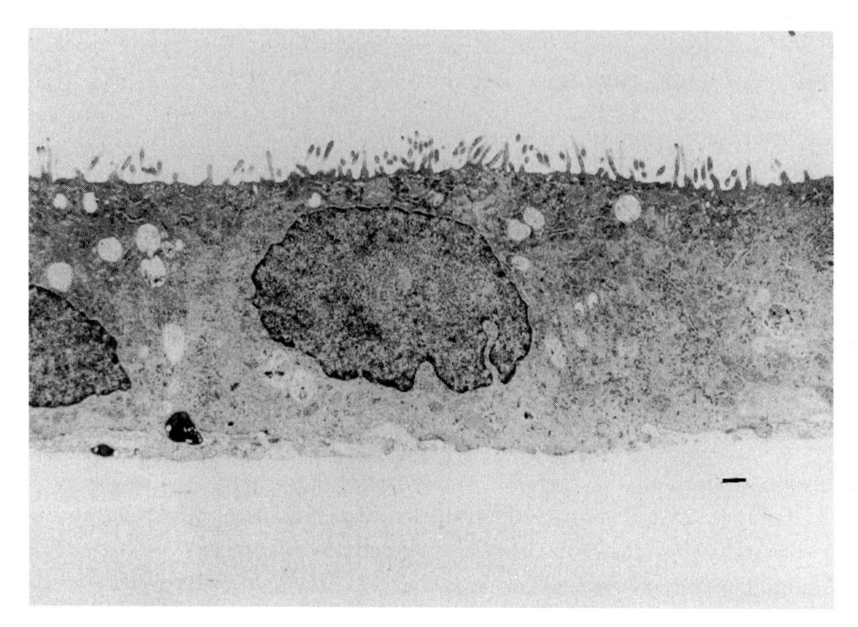

Fig. 4.2. Transmission electron micrograph of a 7-day-old proximal tubule monolayer. The cells are of uniform appearance and show dense apical projections. Scale bar represents 1 μm.

stimulated by PTH but not by vasopressin or isoproterenol (Bello-Reuss & Weber, 1986).

2 The distal nephron monolayers have a high transepithelial resistance (>1 $\Omega \cdot cm^2$) and contain cells resembling either principal or intercalated cells (Fig. 4.3). The principal cell types possess an amiloride-sensitive apical membrane conductance and equivalent short-circuit current (White *et al.*, 1992). Cyclic AMP production is stimulated by vasopressin but not by PTH. The cultures absorb sodium and secrete potassium and these fluxes are stimulated by aldosterone. Cells of these monolayers express a variety of ion channels including amiloride-sensitive Na^+ and barium-sensitive K^+ channels (Ling *et al.*, 1991*a*, *b*) – properties consistent with those of principal cells of the CCD. The monolayers also express intercalated cell function (transport of H^+ and HCO_3^-) and this is also modulated by aldosterone (White *et al.*, 1992).

Longer-term culture and passage

We have used cultures for up to 14 days. Although they continue to grow for several weeks, the electrical resistance falls as the cells overgrow. Both

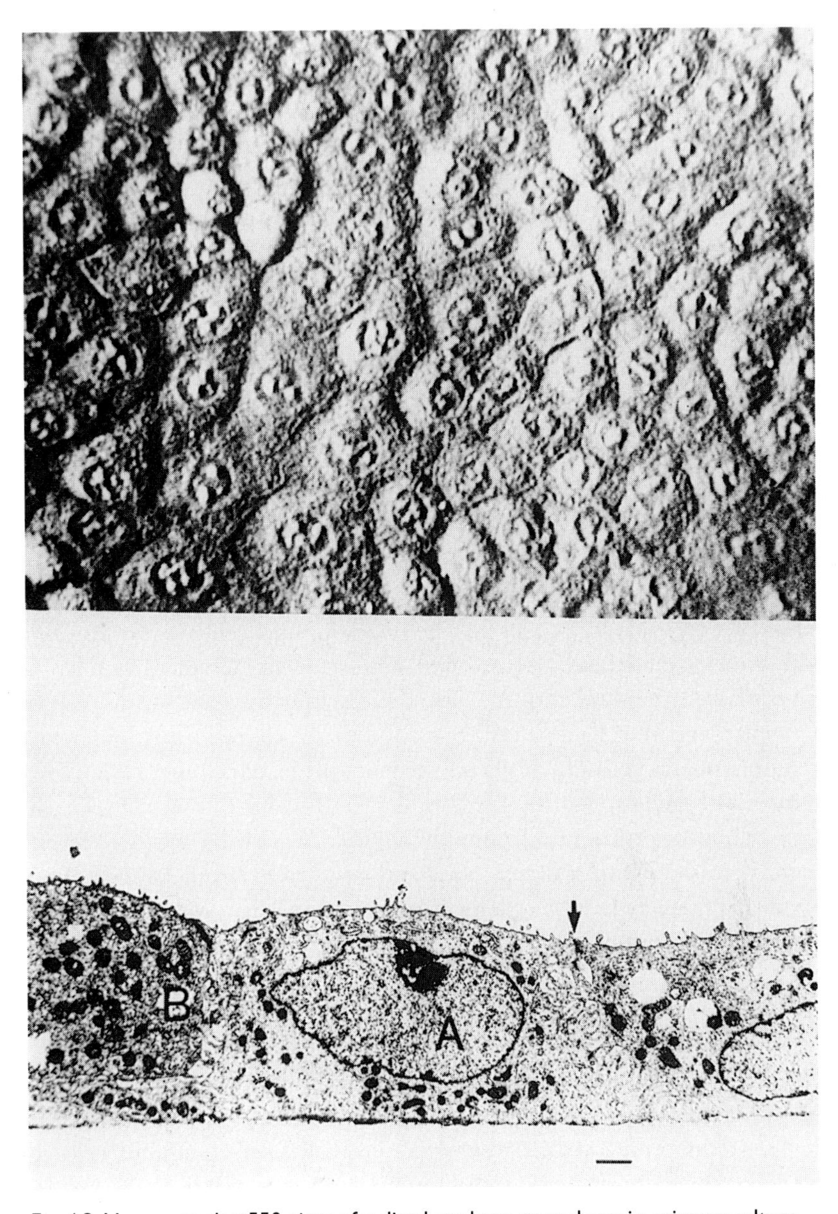

Fig. 4.3. Upper panel: ×550 view of a distal nephron monolayer in primary culture. Lower panel: Transmission electron micrograph of distal nephron monolayer after 7 days in primary culture. Two cell types can be seen: light (the principal cells, A) and dark (the intercalated cells, B).

cultures grow on plastic as well as permeable supports and the CCD cultures have been successfully passaged three or four times with standard enzymatic or non-enzymatic techniques. Moreover, the passaged cells maintain their function. The CCD cells can be cryopreserved by any standard method (e.g. frozen at −1 °C/min in fetal calf serum (FCS) with 10% dimethylsulphoxide). We have not attempted to passage or cryopreserve PT cell cultures.

Culture of bladder epithelium

Structure of the urothelium

The human bladder is lined by a transitional epithelium three to seven cells thick, known as urothelium. Similar cells line the renal pelvis, ureter and urethra proximal to the bladder. Within the urothelium three cell layers follow a unique programme of differentiation: (i) the basal layer consists of columnar/cuboidal cells; (ii) the intermediate layers consist of cuboidal/polypoidal cells, and (iii) the upper luminal cell layer consists of a distinct layer of flattened cells, often binucleate or polyploid, known as 'umbrella cells' because of their shape. Stem cells in the basal layer divide to repopulate the intermediate cell layers, and these cells terminally differentiate into umbrella cells and finally are shed into the bladder lumen. It has been estimated that the loss of umbrella cells is 0.01–0.1% of the total per day. Despite this slow turnover, the urothelium has a remarkable ability to regenerate. For example, if a mouse bladder is stripped of urothelial cells using trypsin, a complete layer of cells regenerates within 3 days and large numbers of mitotic cells can be seen.

Umbrella cells

The highly specialised umbrella cells form a barrier to protect themselves and the rest of the body from the toxic and osmotic influences of urine. In addition to desmosomes, which connect all urothelial cells, umbrella cells are connected by tight junctions at the luminal surface. In addition, the luminal surface is lined by a 13–19 nm thick asymmetrical unit membrane consisting of an outer thick electron-dense layer separated from an inner electron-dense layer by an electron-lucent layer (Jost *et al.*, 1989). Subunits of this membrane are arranged in a lattice, and each subunit is separated by a thinner layer of membrane which appears to act as a hinge. Thus, the whole of the membrane makes a semi-rigid structure which can expand as the bladder distends. Four proteins of the asymmetrical unit membrane have been identified (15, 27, 28 and 47 kDa: uroplakins II, Ia, Ib and III respectively) and antibodies developed (Yo *et al.*, 1994; Lin *et al.*, 1994). The genes for uroplakins share sequence homology with a family of membrane proteins called the 'four transmembrane domain' proteins, which include certain leucocyte differentiation markers. The umbrella cells express cytokeratin 20, which is only found elsewhere in Merkel cells and the epithelium of the gastrointestinal tract. The umbrella cells are also covered by a thick glycocalyx. In addition to contributing to the impermeability of the bladder, this layer may also act as an antibacterial coat (Shupp Byrne *et al.*, 1994).

Protocols for primary culture of urothelial cells

Sources of cells

Urine. Cells can sometimes be cultured from urine samples, although they may be of renal origin. The urine should be collected fresh in a sterile container, centrifuged at 1000 *g* and the cells plated as described below for enzymatically dissociated urothelial cells.

Bladder washings. Irrigating the bladder vigorously with saline (as done to obtain cytology specimens from bladder washes) produces large numbers of urothelial cells, but is uncomfortable unless the donor is anaesthetised. The washes should be transferred to a sterile container and immediately centrifuged at 1000 *g* and plated as described below for enzymatically dissociated urothelial cells.

Surgical biopsies. Tissue can be obtained during reconstructive surgery of the urogenital tract, renal transplantation or cystoscopy. Larger quantities of tissue may be obtained at cystectomy, but this operation is mainly restricted to patients with advanced cancer in whom the urothelium may be abnormal, and the patient may have been treated with radiation or intravesically with chemotherapeutic drugs. Urothelial cells cultured from renal pelvis, ureter or proximal urethra have a similar morphology *in vitro* to those cultured from bladder tissue.

Tissue collection

Materials
Sterile 30 ml plastic Universals

Solutions and reagents

RPMI 1640 medium supplemented with:
L-glutamine 2 mM
Penicillin 10^5 U/l
Streptomycin 100 mg/l
Immediately the biopsy is obtained it should be transferred to the laboratory for processing in the above medium. However, primary cultures can be obtained from biopsies kept at 4 °C for as long as 24 h.

Methods of isolation

Explant culture

Materials

Sterile 10 cm plastic petri dishes
One pair of sterile scalpel handles with Swann Morton no. 11 blades
T25 tissue culture flasks

Solutions and reagents

RPMI 1640 medium supplemented as above
Fetal calf serum (FCS)

Note that batches of FCS differ in their ability to support cell growth. For reproducibility, it is esssential to batch-test sera in advance using 'demanding' human cell lines. Tissue culture suppliers will provide small quantities of one or two batches of FCS free for testing. When the best serum has been selected a large batch can be purchased. It is not possible to predict the suitability of serum on the basis of either price or supplier, and a batch of serum that proves excellent for the growth of cells in one laboratory can be toxic to different cells elsewhere.

Protocol

1 Dice the tissue into 1 mm^3 pieces using crossed scalpel blades.
2 Using a pasteur pipette containing a drop of RPMI 1640 medium, transfer 10–15 explants one at a time in a drop of medium to the bottom surface of a sterile T25 tissue culture flask.
3 Invert the flask and leave the explants to attach for 45 min at 37 °C.
4 Place the flask the correct way up and very gently cover the explants with 5 ml of complete medium (RPMI 1640 supplemented with 2 mM L-glutamine and 10% FCS).

Explant culture is the simplest method for producing primary cultures of urothelium. Within 24 h cells may be seen migrating from the cut edges of the explants and within 14 days a confluent monolayer will be produced. Explant culture has two disadvantages compared with enzymatic dissociation. First, it takes longer to produce a confluent monolayer, and secondly there is a higher incidence of fibroblast contamination, which may only become apparent following subculture.

Enzymatic dissociation

This protocol is based on Petzoldt *et al.*, (1994).

Materials

Dissecting microscope
Sterile 10 cm plastic dishes
Sterile fine iridectomy scissors and watchmaker's forceps
One pair of sterile scalpel handles with Swann Morton no. 11 blades
T25 tissue culture flasks
Coarse sieve (0.5 mm diameter pore size)

Solutions and reagents

0.25% trypsin in 0.2% EDTA (TE)
DMEM with 10% FCS
DMEM and Ham's F-12 medium in a 3:1 mixture supplemented with:
10% FCS
Adenine 1.8×10^{-4} M
Cholera toxin 10×10^{-10} M
Epidermal growth factor 20 μg/l
Hydrocortisone 0.4 mg/l
Liothyronine 2×10^{-11} M
Transferrin 5 mg/l

Protocol

1 Under a dissecting microscope, remove as much stroma as possible.
2 Dice the urothelium into approximately 1 mm^3 pieces using crossed scalpels in a 10 cm plastic petri dish.
3 Incubate in 10 ml of TE for 30 min at 37 °C.
4 Filter the tissue through the coarse (40 mesh) sieve, collecting urothelial cells in DMEM with 10% FCS.
5 Repeat steps 3 and 4. These steps can be repeated up to three times, but with increasing risk of contamination by stromal cells.

6 Collect urothelial cells by centrifugation at 1000 *g* for 5 min.
7 Plate cells in a 3:1 mixture of Ham's F-12 and DMEM supplemented as described above in a T25 flask with 5×10^5 lethally irradiated Swiss 3T3 feeder cells.

An alternative enzymatic method not requiring the use of 3T3 cells has been described by Southgate and co-workers (1994). Using this method, primary cultures can be grown in keratinocyte serum-free medium (which contains bovine pituitary extract rather than serum). Fibroblast growth can be inhibited by the use of feeder cells (e.g. lethally irradiated 3T3 cells) or by using the serum-free medium developed for keratinocytes.

Lethally irradiated Swiss 3T3 cells

Materials

Gamma radiation source
Swiss 3T3 cells (ECACC/ATCC)

Solutions and reagents

DMEM
FCS

Protocol

1 Grow Swiss 3T3 cells in DMEM supplemented with 10% FCS.
2 Passage twice weekly when approximately 90% confluent.
3 Trypsinise cells, suspend in medium without serum and irradiate (6000 rads gamma-irradiation).
4 Wash cells in complete medium, resuspend in DMEM with 10% FCS, count, store at 4 °C and use within 24 h.

Establishment and maintenance of cultures

Almost all biopsies will produce primary cultures (Fig. 4.4). About 5–10% of the primary cultures will be lost due to bacterial contamination, which usually can be traced to an infection in the patient's bladder. Epidermal growth factor and cholera toxin can stimulate cell growth in serum-free or low serum concentrations. Calcium levels are important. At low concentrations (<0.1 mM) proliferation is favoured, but as calcium levels are raised, mitotic activity decreases and at 0.5 mM calcium the labelling index may be reduced by up to 95%. Moreover, the cultures can become multilayered in

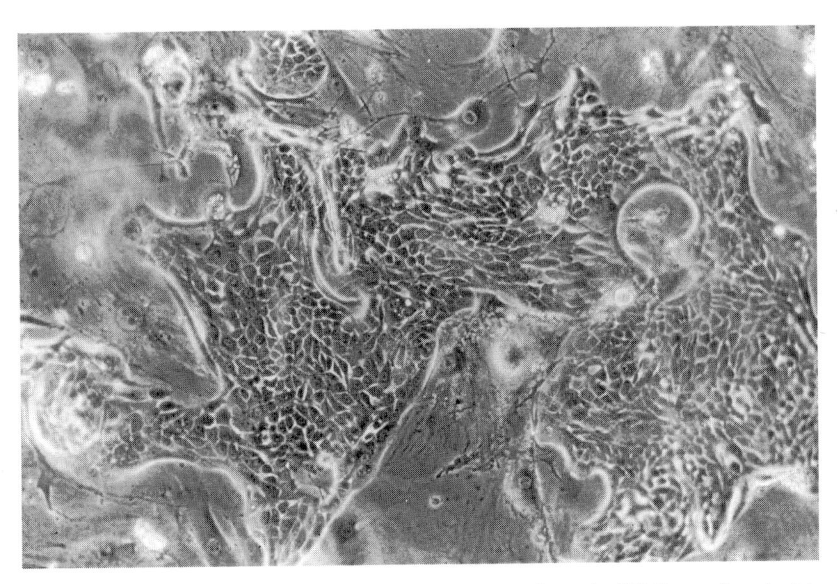

Fig. 4.4. Micrograph of human urothelium in primary culture (×10). Reproduced with permission from Petzoldt *et al.* (1994).

high calcium concentrations. Similar multilayering can be achieved by allowing the cells to continue growing in the same dish after they have become confluent.

Characterisation

Cytokeratins (CK) 7, 8, 17, 18 and 19 are present throughout the urothelium *in vivo* and in all cells growing *in vitro*. All urothelial cells *in vivo* and *in vitro* are negative for CK 4 and 10. Cytokeratin 13 is present in the basal cells and some intermediate cells *in vivo*, and is observed in about 25% of the urothelial cells growing *in vitro*. Cytokeratins 14 and 16 are not expressed *in vivo*, but *in vitro* about 25% of the cells express CK14 and all the cells express CK16. CK16 is characteristic of hyperproliferative conditions, and its expression *in vitro* may reflect the release of the cells from normal growth-regulated conditions or a tendency towards squamous metaplasia. Some studies have also observed CK6 (the partner to CK16) expression in cultured urothelial cells. CK20 is a recently identified cytokeratin, restricted to the urothelium, gastrointestinal tract and Merkel cells, but in the only study which has attempted to demonstrate this antigen in cultured urothelial cells (Southgate *et al.*, 1994), CK20 was not found. Urothelial cells are negative for smooth muscle actin, but vimentin is often expressed by epithelial cells in culture and is therefore a poor negative control. The cells have a normal karyotype.

Human urothelial cells growing *in vitro* do not produce an asymmetrical unit membrane or express the characteristic antigens of umbrella cells, even when growing as a multilayer. There is, however, some evidence that bovine and rodent cells can express uroplakins in culture. When cultures of human urothelium are transplanted to nude mice, a morphology similar to that *in vivo* can be generated, indicating that the cells growing *in vitro* retain the potential for terminal cell differentiation.

Longer-term culture and passage

Depending on the culture conditions and the initial plating density, confluent cultures can be obtained within 1–4 weeks. The cells grow as densely packed small polygonal cells or as more diffuse long fusiform cells. The two cell types are interchangeable and the keratin staining patterns identical. The cultures can be split up to nine times over a 2–3 month period using 0.05% trypsin in 0.02% EDTA. As the cells senesce they become larger and develop a more ragged appearance, and eventually fail to settle and divide following subculture. The cultures can be cryopreserved using standard protocols (e.g. frozen at −1 °C/min in RPMI 1640 supplemented with 50% FCS and 10% dimethylsulphoxide). If serum-free media are being used it will be necessary to incubate the detached cells in trypsin inhibitor before replating.

Urothelial cells can be immortalised with SV40 genes to provide continuous cell cultures. However, most cell lines immortalised with SV40 genes show, with increasing passage, loss of differentiated characteristics and progressive genetic changes (Reznikoff *et al.*, 1993). These changes might be minimised by using a temperature-sensitive construct to immortalise the cells conditionally, but as yet no urothelial cell lines immortalised in this way are available.

An alternative to primary culture of epithelial cells in monolayer is the maintenance of bladder explants in organ culture (i.e. as three-dimensional explant cultures maintained at the fluid/air interface floating on a raft). In serum-containing media, rat bladder has been maintained for up to 160 days and human bladder for up to 33 days.

Acknowledgements

A.S.W. is supported by the National Kidney Research Fund.

References

Bello-Reuss, E. & Weber, M.R. (1986). Electrophysiological studies on primary cultures of proximal tubule cells. *Am. J. Physiol.*, **251**, F490–8.

Fejes-Toth, G. & Naray-Fejes-Toth, A. (1991). Fluorescence activated cell sorting of principal and intercalated cells of the renal collecting duct. *J. Tissue Cult. Methods*, **13**, 173–8.

Fejes-Toth, G. & Naray-Fejes-Toth, A. (1992). Differentiation of renal β-intercalated cells to α-intercalated and principal cells in culture. *Proc. Natl. Acad. Sci. USA*, **89**, 5487–91.

Humes, H.D. & Cieslinski, D.A. (1992). Interaction between growth factors and retinoic acid in the induction of kidney tubulogenesis in organ culture. *Exp. Cell Res.*, **210**, 8–15.

Jost, S.P., Gosling, J.A., Dixon, J.S. (1989). The morphology of normal human bladder urothelium. *J. Anat.*, **167**, 103–15.

Kerjaschki, D., Poczewski, H., Dekan, G., Horvat, R., Balzar, E., Kraft, N. & Atkins, R.C. (1986). Identification of a major sialoprotein in the glycocalyx of human visceral glomerular epithelial cells. *J. Clin. Invest.*, **78**, 1142–9.

Kreisberg, J.I. & Wilson, P.D. (1988). Renal cell culture. *J. Electron Miscrosc. Tech.*, **9**, 235–63.

Lin, J.-H., Wu, X.-R., Kreibich, G. & Sun, T.-T. (1994). Precursor sequence, processing, and urothelium-specific expression of a major 15-kDa protein subunit of asymmetric unit membrane. *J. Biol. Chem.*, **269**, 1775–84.

Ling, B.N., Hinton, C.F. & Eaton, D.C. (1991*a*). Potassium permeable channels in primary cultures of rabbit cortical collecting tubule. *Kidney Int.*, **40**, 441–52.

Ling, B.N., Hinton, C.F. & Eaton, D.C. (1991*b*). Amiloride-sensitive sodium channels in rabbit cortical collecting tubule primary cultures. *Am. J. Physiol.*, **261**, F933–44.

McAteer, J.A., Kempson, S.A. & Evan, A.P. (1991). Culture of human renal cortex epithelial cells. *J. Tissue Cult. Methods*, **13**, 143–8.

Merot, J., Bidet, M., Gachot, B., Le Maout, S., Koechlin, N., Tauc, M. & Poujeol, P. (1989). Electrical properties of rabbit early distal convoluted tubule in primary culture. *Am. J. Physiol.*, **257**, F288–99.

Minuth, W.W., Fietzek, W., Kloth, S., Aigner, J., Herter, P., Rockl, W., Kubitza, M., Stockl, G. & Dermietzel, R. (1994). Aldosterone modulates PNA binding cell isoforms within renal collecting duct epithelium. *Kidney Int.*, **44**, 537–44.

Nosaka, K., Nishi, T., Imaki, H., Suzuki, K., Kuwata, S., Noiri, E., Aizawa, C. & Kurokawa, K. (1993). Permeable type I collagen membrane promotes glomerular epithelial cell growth in culture. *Kidney Int.*, **43**, 470–8.

Petzoldt, J.L., Leigh, I.M., Duffy, P.G. & Masters, J.R.W. (1994). Culture and characterization of human urothelium *in vivo* and *in vitro*. *Urol. Res.*, **22**, 67–74.

Prie, D., Ronco, P., Baudouin, B., Geniteau-Legendre, M., Antoine, M., Piedagnel,

R., Estrade, S., Lelongt, B., Verroust, P.J., Cassingena, R. & Vandevalle, A. (1991). Activation of Simian Virus 40 (SV40) genome abrogates sensitivity to AVP in a rabbit collecting tubule cell line by repressing membrane expression of AVP receptors. *J. Cell Biol.*, **113**, 951–62.

Racusen, L.C., Fivush, B.A., Andersson, H. & Gahl, W.A. (1991). Culture of renal tubular cells from the urine of patients with nephropathic cystinosis. *J. Am. Soc. Nephrol.*, **1**, 1028–33.

Reznikoff, C.A., Kao, C., Messing, E.M., Newton, M. & Swaminathan, S. (1993). A molecular genetic model of human bladder carcinogenesis. *Semin. Cancer Biol.*, **4**, 143–52.

Romero, M.F., Douglas, J.D., Eckert, R.L., Hopfer, U. & Jacobberger, J.W. (1992). Development and characterization of rabbit proximal tubular epithelial cell lines. *Kidney Int.*, **42**, 1130–44.

Shupp Byrne, D.E., MacPhee, M., Mulholland, M., McCue, P., Callahan, H.J. & Mulholland, S.G. (1994). Urinary tract glycoprotein: distribution and antigenic specificity. *World J. Urol.*, **12**, 21–6.

Southgate, J., Hutton, K.A.R., Thomas, D.F.M. & Trejdosiewicz, L.K. (1994). Normal human urothelial cells *in vitro*: proliferation and induction of stratification. *Lab. Invest.*, **71**, 583–94.

Weinstein, T., Cameron, R. & Silverman, M. (1992). Rat glomerular epithelial cells in culture express characteristics of parietal, not visceral, epithelium. *J. Am. Soc. Nephrol.*, **3**, 1279–87.

White, S.J., Boupaep, E.L. & Giebisch, G. (1992). Differentiated transport properties of primary cultured distal nephron. *Cell Physiol. Biochem.*, **2**, 323–35.

Wilson, P.D. (1991). Monolayer cultures of microdissected renal tubule epithelial segments. *J. Tissue Cult. Methods*, **13**, 137–42.

Woo, D.D.L., Miao, S., Pelayo, J. & Woolf, A.S. (1994). Taxol inhibits congenital polycystic kidney disease progression. *Nature*, **368**, 750–3.

Woolf, A.S., Kolatsi-Jouannou, M., Hardman, P., Andermarcher, E., Moorby, C., Fine, L.G., Jat, P.S., Noble, M.D. & Gherardi, E. (1995). Roles of hepatocyte growth factor/scatter factor and Met in early development of the metanephros. *J. Cell Biol.*, **128**, 171–84.

Yo, J., Lin, J.-H., Wu, X.-R. & Sun, T.T. (1994). Uroplakins Ia and Ib, two major differentiation products of bladder epithelium, belong to a family of four transmembrane domain (4TM) proteins. *J. Cell Biol.*, **125**, 171–82.

5

The epididymal epithelial cell

Hsiao Chang Chan and Patrick Y.D. Wong

Introduction

The epididymis is an integral part of the male reproductive system, which, together with the efferent duct and vas deferens, forms the testicular excurrent duct system. The epididymis does not merely serve as a conduit for sperm passage since spermatozoa leaving the testis gradually acquire their motility and fertilising capacity during transit through the epididymis (review by Robaire & Hermo, 1988). Early histological studies on the epididymis revealed its over-elaborate structural features, suggesting that it might have complex functions (Reid & Cleland, 1957). The proposed functions of the epididymis include: (1) transport of spermatozoa, (2) concentration of spermatozoa, (3) maturation of spermatozoa, and (4) storage of spermatozoa.

It is believed that the epididymis creates a specific and optimal microenvironment for sperm maturation and storage through the absorptive and secretory activities of its epithelial lining. In mammals, the epididymis absorbs more than 95% of the fluid coming from the testis. The absorption of macromolecules and particulate material, mostly via pinocytosis, from the epididymal lumen has been demonstrated throughout the epididymis. On the other hand, the epididymal epithelium also actively transports substances into the lumen such as lipids, steroids, enzymes and glycoproteins. Through its absorptive and secretory activities, the epididymal epithelium alters the contents of the luminal fluid, which differs from rete testis fluid and blood plasma in the concentration of various ions, enzymes and organic components.

Although the epididymal epithelium is thought to be essential for sperm maturation, its precise role, especially with regard to the functions of particular cell types, is largely unknown. Attempts have been made to develop methods for the isolation of viable epididymal epithelial cells in order to obtain more specific information about the functions of the epididymis (review by Robaire & Hermo, 1988). The success in culturing epididymal epithelial cells has enabled direct evaluation of epididymal epithelial cell functions including studies on the following aspects: (1) absorptive and secretory activities of epididymal epithelial cell populations; (2) steroidogenesis by specific cell types; (3) hormonal regulation of different epididymal cell types; (4) regional functions of epididymal cell populations; (5) sperm–epididymis interaction, by co-culturing sperm with epithelial cells.

The importance of the epididymal epithelial cell culture is also emphasised by the fact that the epithelium lining the epididymis is directly involved in the pathology of cystic fibrosis (CF), the most common lethal autosomal recessive genetic disease in Caucasian peoples. More than 97% of male CF patients are sterile due to abnormalities in the epididymis and to partial or complete absence of the vas deferens. The basic defect of CF is thought to be in the regulation of Cl^- transport. Defective electrolyte and fluid secretion across the epithelium of the epididymis could lead to obstructive azoospermia as seen in Young's syndrome and some other cases of unexplained male infertility, as well as in CF. Therefore, epididymal epithelial cell culture has been applied to electrophysiological studies in an attempt to investigate the mechanisms underlying epididymal electrolyte and fluid transport and its regulation.

Using the patch-clamp technique, we have characterised different types of Cl^- channels in cultured rat epididymal cells (Huang *et al.*, 1993). Parallel studies on cultured human fetal epididymal cells have also been carried out (Pollard *et al.*, 1991). ATP-activated cation channels, in addition to Cl^- channels, have been identified in cultured human epididymal cells (Chan *et al.*, 1995*b*). The presence of several adrenergic receptors has also been demonstrated by examining the effect of various adrenergic agonists and antagonists on whole-cell Cl^- currents in cultured rat epididymal cells (Chan *et al.*, 1994). In this study, different adrenergic receptor subtypes were found to be present in single rat epididymal cells, which may imply the importance of adrenergic regulation in fine tuning of the secretory process in the epididymis.

However, conventional cell culture systems do not permit investigation of the polarised functions, such as epididymal secretion, of epithelial cells. Demonstration of transepithelial electrolyte transport and vectorial secretion by epithelial cells has been made possible by the advent of permeable supports for monolayer culture of epididymal epithelial cells. Active anion secretion was first demonstrated in reconstituted rat epididymal epithelium in culture using the short-circuit current technique (Cuthbert & Wong, 1986). The anion secretion across the epididymal epithelium has been shown to be under adrenergic influence even in culture conditions; for example, the short-circuit current measured from cultured monolayers could be stimulated by exogenous adrenaline (Wong, 1988). In cultured monolayers obtained from various regions of both rat and human epididymides, we have been able to demonstrate regional differences in bioelectrical properties and hormonal regulation along the epididymal tract (Chan *et al.*, 1995*a*). Extensive analysis of neurohumoral and local control of epididymal anion secretion has also been made possible by this culture technique (review by Wong *et al.*, 1992). Epididymal cells from CF transgenic mice (Snouwaert *et al.*, 1992), a mouse model of CF, have also been cultured to form monolayers; these exhibited defective cAMP-regulated, but not Ca^{2+}-regulated, anion secretion (A.Y.H. Leung *et al.*, unpublished data). The studies with cultured epididymal cells may enable us to gain better understanding of epididymal functions in relation to the disease of CF and other cases of male infertility.

In addition to the studies of epididymal functions, another potential application of the epididymal cell culture is for rapid screening of antifertility drugs, since the epididymis has been suggested to be a possible site of action for a male contraceptive (review by Cooper, 1992). The epididymis is thought to be important for sperm maturation by creating an optimal microenvironment on which transformation of an immature sperm to one capable of movement and fertilisation depends. It is this indispensable role of the epididymis that has given rise to the 'epididymal approach', an antifertility strategy based on intervention in sperm function at a post-testicular level. Contraceptives could interfere with epididymal cell functions, such as inhibiting secretory protein synthesis or transport activities, thereby creating a hostile environment for epididymal spermatozoa. For example, treatment of rats with reserpine, a compound that depletes cellular noradrenaline, serotonin and dopamine reserves, decreases fluid secretion in the epididymis by inhibiting Cl^- secretion, which leads to increased luminal fluid viscosity and sperm concentration (Wen & Wong, 1988). Reserpine-induced functional disturbances in the epididymis resemble those produced by CF, making it a potential male contraceptive agent. The combination of the epididymal culture technique and electrophysiological methods proves to be practical and advantageous for examining the effect of contraceptives on epididymal electrolyte transport.

Considering the wide range of potential applications of the epididymal epithelial culture, we describe here a culture technique currently used in our laboratory which has proved to be useful for a variety of research purposes. Although the method emphasises the particulars for culture on permeable supports, it may be used to establish conventional monolayer cell culture systems as well. The reader is referred to the review by Robaire & Hermo (1988) for the historical development and detailed characterisation of epididymal cell culture.

Epithelial cell types in the epididymis

A number of schemes have been proposed for dividing the epididymis into different segments. In this chapter, we view the epididymis as being divided into three anatomical segments – head (caput), body (corpus) and tail (cauda) – as shown in Fig. 5.1 in relation to other parts of the male reproductive system (the testis, efferent duct and vas deferens). An initial segment which lies between the efferent ducts and the caput epididymidis is also illustrated.

The regional changes in the histology of the epididymis have been described for the rat (Reid & Cleland, 1957). Five distinct types of cells are found in the epididymal epithelium: principal, narrow, clear, basal and halo cells. The striking features of principal cells include numerous stereocilia, apical vesicles, vacuoles and multivesicular bodies, an extensive supranuclear Golgi apparatus, and perinuclear and basal granular endoplasmic reticulum. The basal cells are located adjacent to the basal lamina of the epithelium; they have a dense triangular nucleus and very few cytoplasmic organelles. The halo cells are small and round with a dense nucleus and a pale cytoplasm with sparse organelles. The clear cell contains numerous electron-lucent apical vesicles, vacuoles and a lysozyme-like structure that contains dense materials.

The distribution of cell types varies along the epididymal epithelium. The principal cell is the most abundant cell type of the five; it contributes 80% of the total epithelial cells in the initial segment, the proportion gradually decreasing to 65% in the cauda epididymidis. Transportation (of electrolytes, proteins, enzymes and hormones) is thought to be the main function of the principal cell. Ultrastructural and autoradiographic studies indicate that principal cells are actively involved in protein and glycoprotein secretory activities and, to a lesser extent, the absorption of fluid and particulate material. Therefore, efforts have been made to develop cultures that are enriched in the principal cell population for use in studies of secretion and reabsorption by the epididymal epithelium.

Source of cells

Epididymal epithelial cell cultures have been obtained from many different species including the rat, mouse, rabbit, sheep, cow, and human. The rat epi-

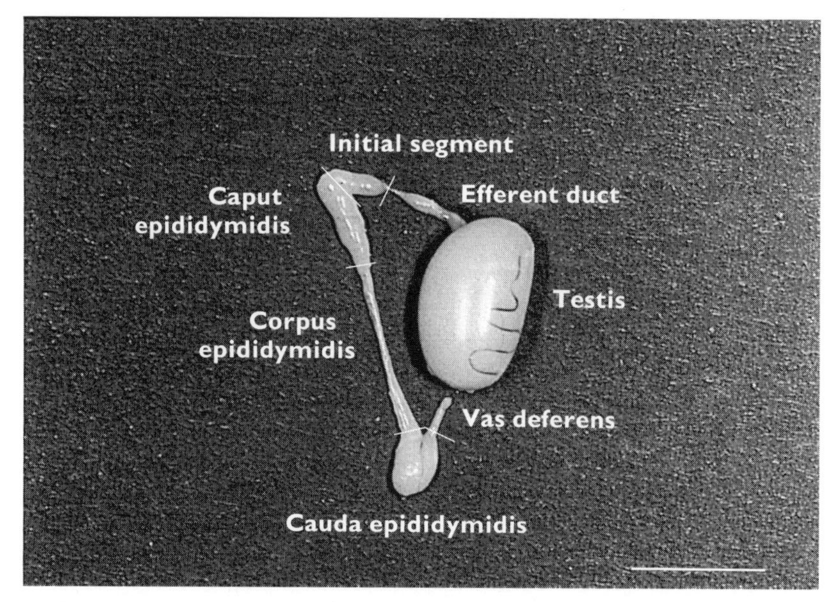

Fig. 5.1. Illustration of the male reproductive system of the rat: testis, efferent duct, epididymis and vas deferens. The epididymis is further divided into three regions: caput (head), corpus (body) and cauda (tail). An initial segment which lies between the efferent duct and caput epididymidis is also illustrated. Scale bar represents 1 cm.

didymal culture is by far the most extensively studied. Epididymal epithelial cells (about 80% principal cells) isolated by various techniques, from rats of different ages, have been reported to grow in culture (review by Robaire & Hermo, 1988). However, pubertal rats (about 40 days old) are preferred as a source of cells for culture on permeable supports since it is more difficult to achieve consistent confluent monolayer formation when adult rats are used, possibly due to the contamination by spermatozoa. The requirement for consistent confluent monolayers is imposed by the subsequent electrophysiological studies; for example, the transepithelial resistance of a 'leaky' monolayer would be too small to generate meaningful short-circuit current measurements.

The function of different regions along the male excurrent duct may differ between species and thus information about the epididymal regions of one species may not necessarily translate well to another species. Therefore, it is necessary to obtain information from human tissues in order fully to understand the functions of the human epididymis and to help advances in the treatment of male infertility or the design of post-testicular contraceptives. The limitation of experimentation on humans makes it even more important to develop an epididymal culture of human origin. The source of human epididymis is rather limited. Most human epididymides are obtained from patients, usually elderly men, undergoing orchidectomy for prostatic cancer (Cooper *et al.*, 1990; Chan *et al.*, 1995*a*). Therefore, one should be cautious when interpreting data obtained from these tissues. Another source of human epididymides is fetuses – either mid-trimester prostaglandin-induced terminations or spontaneous abortions (Harris & Coleman, 1989; Coleman & Harris, 1991). It should be kept in mind that while the components of the male reproductive system are anatomically well developed, and can even exhibit CF phenotypes, at that stage of development, the differentiated functions of the epididymis may not be fully expressed. Nevertheless, despite the potential limitations of human epididymal epithelial cultures they are extremely valuable.

Isolation of rat epididymal epithelial cells

Materials

Eagle's Minimum Essential Medium (EMEM; GIBCO)
Fetal calf serum (FCS; GIBCO)
Penicillin (Sigma) 100 U/ml
Streptomycin (Sigma) 100 μg/ml
L-glutamine (GIBCO) 200 mM 100×
5α-Dihydrotestosterone (5αDHT; Sigma, A8380): dilute to 2 μM in HBSS
Hank's Balanced Salt Solution (HBSS; GIBCO)
Trypsin type II (Sigma, T8128) 0.25% (w/v) in 10 ml HBSS
Collagenase type I (Sigma, C0130) 0.1% (w/v) in 10 ml HBSS
Non-essential amino acids 100× (GIBCO)
Sodium pyruvate type II (Sigma, P2256)
Hyaluronidase (Sigma)
Elastase (Sigma)

For 500 ml EMEM add:
5 ml penicillin (100 U/ml)/streptomycin (100 μg/ml)
10 ml L-glutamine (4 mM)
0.25 ml 5αDHT (1 nM)
10 ml FCS 10%
55 mg sodium pyruvate type II (1 mM)
5 ml non-essential amino acid 100×
Petri dish/flask (Falcon)

Materials for permeable supports
Millipore filter HAWP 02500 (Millipore)
Sylgard 184 silicone elastomer kit (Dow Corning)
Collagen type VII (Sigma) 2.5 mg/ml in 0.2% acetic acid

Different methods have been used to isolate epididymal epithelial cells, including unit gravity sedimentation, elutriation, sequential enzyme treatment and Percoll gradient centrifugation (review by Robaire & Hermo, 1988). The current method of choice in our laboratory for the isolation of epididymal epithelial cells from rats is adapted from that developed by Kierszenbaum *et al.* (1981), but with an emphasis on the conditions required for culture on permeable supports. The isolation method is illustrated schematically in Fig. 5.2.

Protocol

1 Place killed Sprague–Dawley rats (male, 200–230 g) on a sterilised workbench and wipe their lower abdomen with 70% ethanol.
2 Make an incision in the lower abdomen near the scrotum and quickly dissect out the testis and the excurrent ducts. Immerse the dissected tissue in sterile Hank's Balanced Salt Solution (HBSS) in a 10 cm petri dish.
3 Transfer the tissue in this petri dish to a laminar flow cabinet. Hold the entire dissected organ complex by the vas deferens with a pair of fine watchmaker's forceps. Next, trim off the fat along the length of the epididymis before cutting the epididymis off the testis. Separate the epididymis from the testis by cutting off the connective tissue between the two, starting from the cauda, using a pair of fine scissors.
4 Different regions of the epididymis may be obtained under a dissecting microscope according to the anatomical division shown in Fig. 5.1. The capsule around the caput epididymidis has to be removed in order to separate the initial segment and the rest of the caput epididymidis.
5 Mince the tissues, now free of fat and connective tissue, finely with a pair of curved iris scissors. It is important that the fragments be small; this procedure may take 10 min and the operator's wrists usually ache if the desired fineness of the tissue is achieved. During this procedure drops of HBSS solution may be applied to the tissue to avoid the tissue drying out. A paste-like mixture, rather than fragments, of tissues should be obtained at the end of this procedure.
6 Transfer the minced tissues to a 25 cm^2 culture flask containing 0.25% trypsin (w/v) in 10 ml HBSS, and incubate the flask for 30 min in a shaking water bath at 32 °C with speed of 160–180 strokes/min.

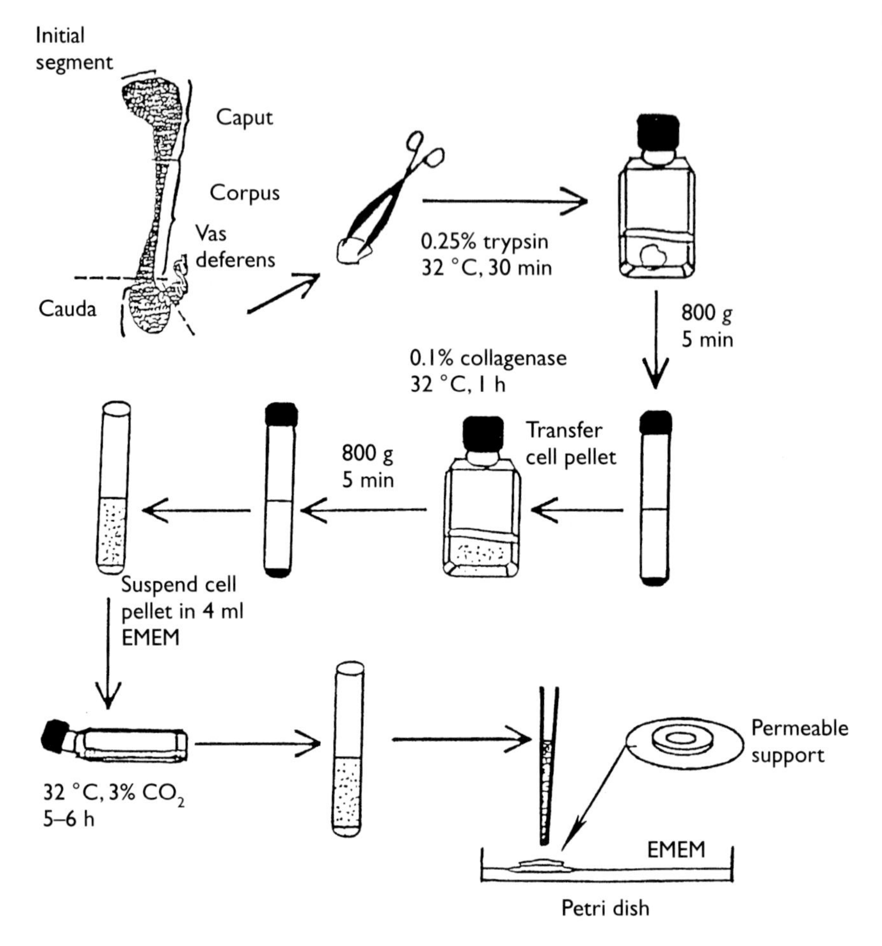

Fig. 5.2. Schematic illustration of the procedure for isolating rat epididymal epithelial cells.

7 Transfer the enzyme-treated tissues from the culture flask to a 15 ml tube and collect the tissue by centrifugation at 800 *g* for 5 min. Discard the supernatant and resuspend the pellet in 10 ml HBSS containing 1 mg/ml collagenase (equivalent to 330 units/mg of solid). Put the tissue suspension into the culture flask again and incubate in the water bath at 32 °C with the same shaking speed, for 60 min.

8 Apply the same centrifugation procedure as in step 7 to separate isolated cells and resuspend the pellet in Eagle's Minimum Essential Medium (EMEM) containing non-essential amino acids (0.1 mM), sodium pyruvate (1 mM), L-glutamate (4 mM), 5α-dihydrotestosterone (1 nM), 10%

fetal bovine serum, penicillin (100 IU/ml) and streptomycin (100 μg/ml). If isolated cells are for culture on permeable supports, cell dilution at this stage is crucial. Since it is not feasible to count cells, due to the presence of cell aggregates, cell pellet volume is measured, and 4 ml of complete medium is added per 200 μl of cell pellet.

9 Transfer the cell suspension into a 25 cm^2 culture flask and place it in an incubator at 32 °C with 5% CO_2 for 5–6 h to remove non-epithelial cells such as fibroblasts and smooth muscle cells, which adhere to the substrate during this time.

10 Collect the epithelial cells suspension from the flask using a pipette and replate for further culture.

Isolation of human epididymal epithelial cells

Human epididymis epithelial cells are isolated using high concentrations of collagenase, hyaluronidase and elastase, on the basis of a method developed by Cooper *et al.* (1990). Elastase is used as the human epididymis contains abundant elastic tissues. High concentrations of enzymes are used to avoid adherence of epithelial fragments to the connective tissue, a phenomenon that is induced by low concentrations of enzymes. The isolation method is illustrated in Fig.5.3.

Protocol

1 Immediately after orchidectomy, transport tissues from the hospital to the laboratory, on ice. Remove the capsule of the human epididymis and separate the tubules from the connective tissues with two pairs of fine forceps using a dissecting microscope. It is more difficult to prepare tissue from the caput and cauda epididymidis as the tiny epididymal tubules in the caput are not readily freed from surrounding connective tissue and the cauda is not usually intact. Therefore only corpus epididymidis tissue is usually obtained for culture, as illustrated in Fig. 5.3.

2 Immerse the tubules in 10 ml HBSS containing 80 mg collagenase (type I, equivalent to 24000 U) and 0.5 mg elastase (equivalent to 38.4 U). Incubate the tubule suspension in a water bath for 15 min at 37 °C with a shaking speed of 120 strokes/min (alternatively, incubate the suspension overnight at 4 °C, without shaking).

3 Place enzyme-treated tubules onto a petri dish and free the tubules from the remaining connective tissues using forceps. Repeat this process with fresh enzyme solution until the tubules are completely freed from connective tissues and blood vessels.

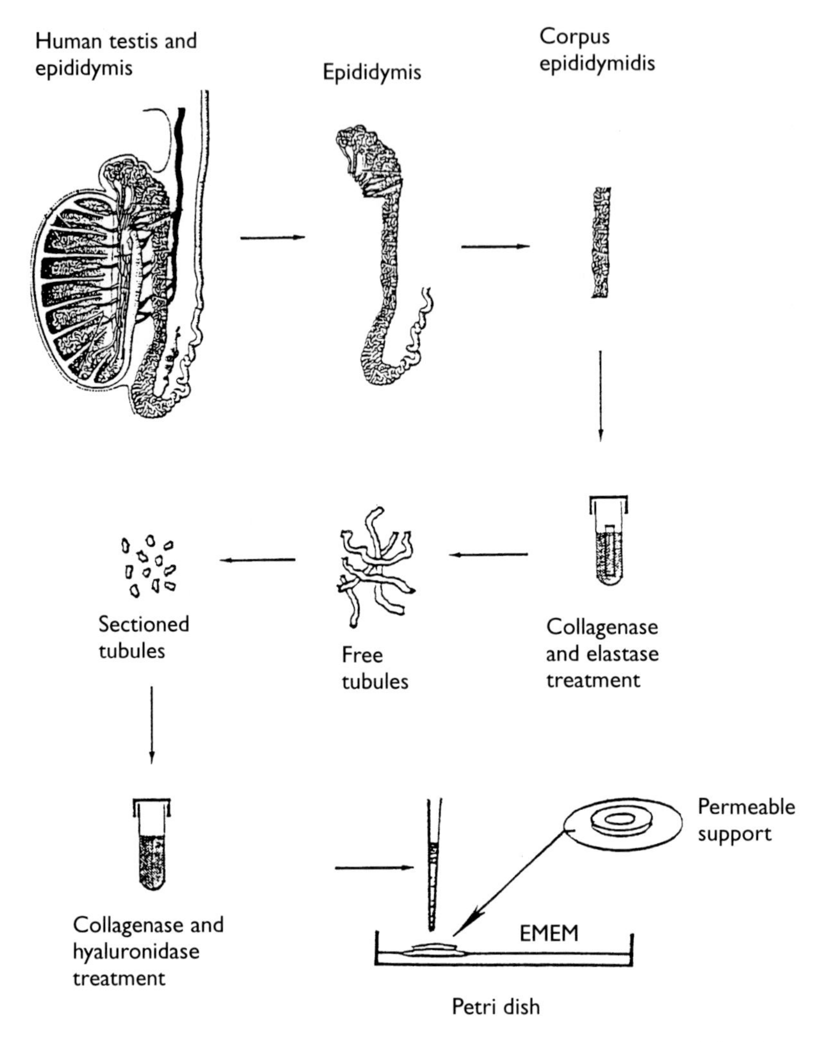

Fig. 5.3. Schematic illustration of the procedure for isolating human epididymal epithelial cells.

4 Mince the tubules with methods similar to that described above for the rat epididymis (step 5, p. 85). However, human tissues can not be processed to the same degree of fineness as that obtained in rat tissues; fragments, rather than a paste-like mixture, of the tubules remain at the end of the procedure.
5 Transfer the tubule fragments into a 25 cm^2 culture flask with 10 ml HBSS containing 24000 U collagenase and 3125 U hyaluronidase. Incubate the

suspension for 45 min in water bath at 37 °C with a shaking speed of 120–150 strokes/min.

6 After incubation, collect isolated cells by centrifugation at 800 *g* for 5 min. Resuspend the pellet in fresh enzyme solution and incubate the cell suspension for an additional 45 min under the same conditions described in step 5 above
7 Wash the cells once with EMEM and resuspend them in the same medium as that used for the culture of rat epididymal cells. For culture on permeable supports, add 8 ml medium to 300 μl of cell pellet.
8 Minimise non-epithelial cell contamination by replating (as described for the rat, p. 85), or by using D-valine medium (Cooper *et al.*, 1990).

Primary cultures

The cells isolated by the procedures described above are ready for culture. Various substrates could be used to establish cultures from this cell preparation: glass coverslips (for histological study), small culture dishes (for patch-clamp study) or 24 well plates (for radioimmunoassay). The method described here is for culture on permeable supports.

Preparation of permeable supports

The permeable support is made of a membrane filter, with a silicone ring attached to confine the area of the cell monolayer (Fig. 5.4). The membrane used is a Millipore filter with a pore size of 0.45 μm, which may be coated with collagen. Place 0.1 ml of collagen solution (type VII, dissolved in 0.2% acetic acid, 2.5 mg/ml) into a 25 mm diameter Millipore filter, spread evenly and allow to dry at room temperature. The permeable supports should be prepared prior to cell isolation.

Protocol

1 Mix silicone elastomer well with curding agent (10:1); both reagents are supplied in the Sylgard 184 silicone elastomer kit (Dow Corning).
2 Weigh 8 g of the mixed silicone resin and pour onto a 10 cm petri dish.
3 Place the silicone resin-containing petri dish on a horizontal apparatus until the silicone resin settles.
4 Put the silicone resin-containing petri dish into a 50 °C oven for 2 h until the silicone resin curds.
5 Punch the cured silicone resin with a double-barrelled cutter to make silicone rings. The inner diameter of the rings may be varied from 0.1 to

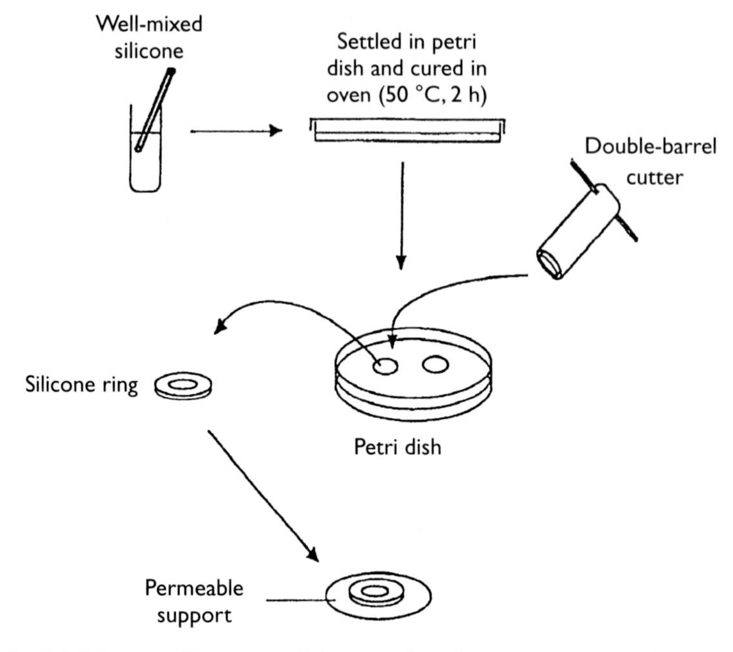

Fig. 5.4. Schematic illustration of the procedure for preparing permeable supports.

0.45 cm^2 so that they can be used selectively depending on available cell numbers.

6 Glue the silicone ring onto the Millipore filter (which may have been coated with collagen) with silicone rubber to form a permeable support.
7 Put the newly made permeable supports on a horizontal apparatus and let them dry at room temperature overnight. It is important that the newly made permeable supports be put on a horizontal plane to avoid deformation.
8 Sterilise the permeable supports by ultraviolet (UV) irradiation overnight.

Culture of rat epididymal epithelial cells

Protocol

1 Fill a 10 cm petri dish with 15 ml complete medium (p. 84). Use a pair of fine forceps to transfer four permeable supports to the medium-filled petri dish and let them float on top of the medium.
2 Vigorously pipette the isolated-cell suspension with a 1 ml pipette to break up clumps. Hold the cell-containing pipette vertically, directly above the

permeable support, and transfer 0.25 ml of cell suspension to each collagen-coated permeable support with a diameter of 0.45 cm. It is important that this procedure be carried out quickly to avoid settling of the cells within the pipette. It is also crucial that this operation be performed steadily to keep cell-containing permeable supports floating; have one hand steadying the pipette and the other one dispensing from it. Unsteady operation may disturb the balance of the permeable support resulting in spill-over of cells.

3 Cover the petri dish, then carefully hold it horizontally and transfer it to a humidified incubator at 32 °C with 5% CO_2.
4 After 2 days, remove medium from the petri dish and re-feed with fresh medium gradually, so that the permeable supports can float on top of the medium again. Under these conditions, cultures normally reaches confluency in 4 days and are ready for electrophysiological studies.

Culture of human epididymal epithelial cells

Protocol

1 Fill a 10 cm petri dish with 15 ml complete medium (p. 84). Use a pair of fine forceps to transfer four permeable supports to the medium-filled petri dish and let them float on top of the medium.
2 As for step 2 above.
3 Incubate cells at 37 °C in a humidified incubator with 5% CO_2.
4 After 2 days, remove medium from the petri dish and re-feed with fresh medium gradually, to keep the permeable supports floating.
5 Change medium every 3 days (same procedure as in previous step 4) until confluency is reached (about 50 days).

Characterisation of the epithelial cells in culture

Morphology

The epididymal epithelial cells in culture may be evaluated morphologically by light or electron microscopy. Principal cells contain dense granules which emit yellow-orange autofluorescence under UV illumination. These granules represent a useful marker for identification of principal cells (Olson *et al.*, 1983). The epithelial nature of the culture can also be evaluated by positive staining with anti-keratin antibodies, since keratins are found only in epithelial cells and not in smooth muscle cells or connective tissue.

Progressive attachment of rat epididymal epithelial cells, obtained by various isolation techniques, to various substrates (glass, plastic or filter membrane) can be seen after 24–48 h in culture with or without collagen or extracellular matrix. The attachment of human epididymal cultures from elderly men may take more than a week. Although the cells seem inactive for some period of time, they do eventually reach confluency as is shown in Fig. 5.5*a*. The cells retain structural features characteristic of epithelial cells in the intact epididymis, such as a well-developed Golgi apparatus, cell junctions, tonofilaments, autofluorescent cytoplasmic granules with a lamellar substructure, and surface microvilli (Olson *et al.*, 1983). However, a change in cell shape, from columnar to flattened-polygonal, and loss of apical microvilli occur under conventional culture conditions. These differentiated characteristics appear to be better maintained when the cells are cultured on permeable supports, as is shown in Fig. 5.5*b*. We note that plating density is an important factor in determining whether highly polarised confluent monolayers can be obtained. Some stacking of cells is often observed when the cultured monolayer exhibits cuboidal and polarised characteristics (Fig. 5.5*b*). A flattened cell appearance is usually associated with low plating density (not shown). It has been observed that the cell height as well as the appearance of the microvillous border is a function of plating density (Byers *et al.*, 1986).

Differentiated functions

The epididymal epithelial cell cultures also express differentiated functions, which may be assessed by a number of techniques. Using radioisotopes, the ability of epididymal epithelial cells to metabolise testosterone to its 5α-reduced metabolites has been demonstrated in rat cultures (Robaire & Hermo, 1988). Principal cells have been shown to synthesise and secrete acidic epididymal glycoprotein, hence the presence of acidic epididymal glycoprotein in cultured rat epididymal epithelial cells and its secretion into the culture medium can be studied by immunocytochemistry and radioimmunoassay, respectively (Kierszenbaum *et al.*, 1981). Expression of certain mucins, known to be present on the luminal surface of adult epididymis, may be further evidence in support of a ductal origin for the cultured cells, as has been shown for cultured human fetal epididymal epithelial cells (Harris & Coleman, 1989). Vectorial absorptive and secretory activities can also be maintained in epididymal culture on permeable supports (Byers *et al.*, 1986). Adsorptive endocytosis from both apical and basolateral aspects, and secretion of alkaline and acid phosphatases and *N*-acetylgucosaminidase have been observed in human epididymal epithelial cells grown on permeable supports (Cooper *et al.*, 1990).

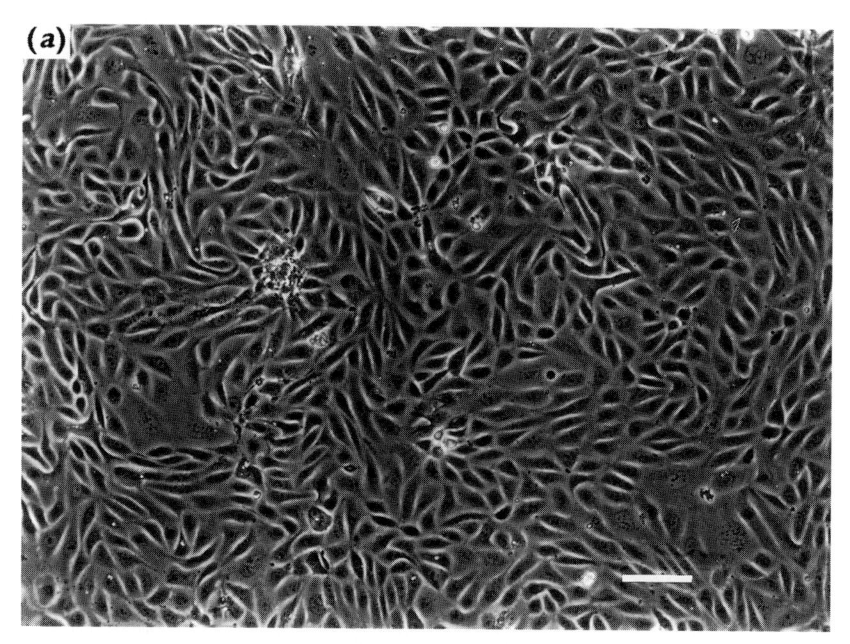

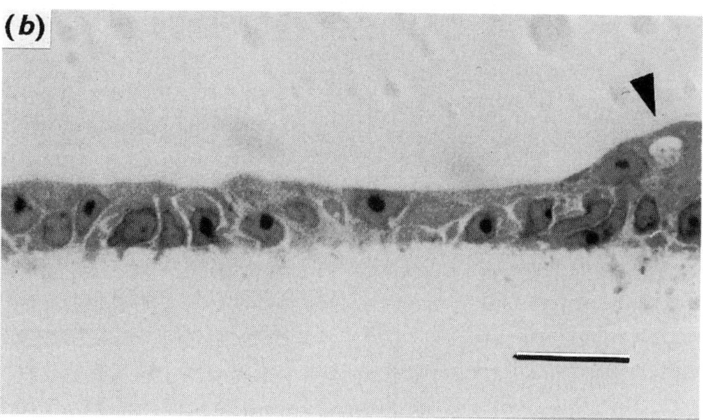

Fig. 5.5. (*a*) Phase contrast micrograph of human corpus epididymal monolayer after 10 days in culture. Cells were grown on a petri dish. Scale bar represents 10 μm. (*b*) Resin section (1 μm thick) of a rat cauda epididymal monolayer grown on a permeable support (Millipore filter). Cells were fixed and processed for microscopy after 4 days in culture. The cells are cuboidal with large nuclei and prominent nucleoli. Some stacking of cells (arrow) is often observed when a cultured monolayer exhibits cuboidal and polarised characteristics. Scale bar represents 20 μm.

Culture on permeable supports also permits routine electrical resistance measurement, for example by the short-circuit current technique, to determine whether a confluent polarised monolayer has been obtained. It has been demonstrated in various culture preparations that transepithelial resistance of culture monolayers increases with the increase in culture time, until it reaches a maximum level and subsequently decreases. Under our culture conditions, a maximum of transepithelial resistance is usually obtained after 4 days and about 50 days in culture for rat and human epididymal epithelial cells, respectively (Chan *et al.*, 1995a). This correlates well with morphological observations, as shown in Fig. 5.5*b* for rat culture, exhibiting confluent and polarised characteristics. Using the short-circuit current technique, electrolyte secretion across cultured epididymal monolayers can be studied. Regional differences in anion secretion have been shown to be maintained in cultured epithelia from rat and human male excurrent ducts (Chan *et al.*, 1995*a*).

Long-term culture

Many laboratories have reported that isolated epididymal epithelial cells have a relatively low rate of [^{3}H]thymidine incorporation (DNA synthesis) and a low mitotic index, suggesting that long-term culture is not feasible. However, subcultures of human fetal epididymal epithelial cells have been reported and cells could be passaged at least five times with 0.25% trypsin and 0.02% EDTA (Harris & Coleman, 1989). Immortalised human epididymal epithelial cell lines have been generated by transformation of primary epithelial cells with a plasmid containing an origin-defective SV40 virus plasmid (Coleman & Harris, 1991). These may be useful for some studies, though it is likely that they have lost or down-regulated a number of markers of differentiation subsequent to the transformation process.

Other considerations

When growing epididymal epithelial cells particular attention should be paid to the following points:

1 Since establishment of these cultures is a long procedure it is essential to maintain sterility throughout. For example, ensure by UV irradiation that the permeable supports are sterile.
2 Use fresh cell culture medium.
3 Since it is not possible to obtain accurate cell counts as the cells are isolated as clumps, cell dilution before plating is done on the basis of the

volume of the cell pellet. This estimation is crucial, though it may take some practice to achieve, since variation in cell density may result in inconsistency in transepithelial resistance measurements.

Acknowledgements

The authors would like to thank Mr W.O. Fu for his useful input and assistance in preparing this manuscript and figures, and Mrs P.Y. Leung for her technical assistance in obtaining micrographs. Some of the work described here was supported by the Research Grants Committee of Hong Kong.

References

Byers, S.W., Hadley, M.A., Djakiew, D. & Dym, M. (1986). Growth and characterization of polarised monolayers of epididymal epithelial cells and sertoli cells in dual environment culture chambers. *J. Androl.*, **7**, 59–68.

Chan, H.C., Fu, W.O., Chung, Y.W., Zhou, T.S. & Wong, P.Y.D. (1994). Adrenergic receptors on cultured rat epididymal cells: regulation of Cl^- conductances. *Biol. Reprod.*, **51**, 1040–5

Chan, H.C., Lai, K.B., Fu, W.O., Chung, Y.W., Chan, P.S.F. & Wong, P.Y.D. (1995*a*). Regional differences in bioelectrical properties and anion secretion in cultured epithelia from rat and human male excurrent ducts. *Biol. Reprod.*, **52**, 192–8.

Chan, H.C., Fu, W.O., Chung, Y.W., Chan, P.S.F. & Wong, P.Y.D. (1995*b*). An ATP-activated cation conductance in human epididymal cells. *Biol. Reprod.* **52**, 645–52.

Coleman, L. & Harris, A. (1991). Immortalization of male genital duct epithelium: an assay system for the cystic fibrosis gene. *J. Cell Sci.*, **98**, 85–9.

Cooper, T.G. (1992). The epididymis as a site of contraceptive attack. In *Spermatogenesis–Fertilization–Contraception: Molecular, Cellular and Endocrine Events in Male Reproduction*, ed. E. Nieschlag & U.-F. Habenicht pp. 419–60. Berlin: Springer-Verlag.

Cooper, T.G., Yeung, C.H., Meyer, R. & Schulze, H. (1990). Maintenance of human epididymal epithelial cell function in monolayer culture. *J. Reprod. Fertil.*, **90**, 81–91.

Cuthbert, A.W. & Wong, P.Y.D. (1986). Electrogenic anion secretion in cultured rat epididymal epithelium. *J. Physiol.* (*Lond*), **378**, 335–45.

Harris, A. & Coleman, L. (1989). Ductal epithelial cells cultured from foetal epididymis and vas deferens: relevance to sterility in cystic fibrosis. *J. Cell Sci.*, **92**, 687–90.

Huang, S.J., Fu, W.O., Chung, Y.W., Zhou, T.S. & Wong, P.Y.D. (1993). Properties of cAMP-dependent and Ca^{2+}-dependent whole cell Cl^- conductances in rat epididymal cells. *Am. J. Physiol.*, **264**, C794–802.

Kierszenbaum, A., Lea, O., Petrusz, P., French, F.S. & Tres, L.L. (1981). Isolation, culture, and immunocytochemical characterization of epididymal epithelial cells from pubertal and adult rats. *Proc. Natl. Acad. Sci. USA*, **78**, 1675–9.

Olson, G.E., Jonas-Davies, J., Hoffman, L.H. & Orgebin-Crist, M.C. (1983). Structure features of culture epithelial cells from the adult rat epididymis. *J. Androl.*, **4**, 347–60.

Pollard, C.E., Harris, A., Coleman, L. & Argent, B.E. (1991). Chloride channels on epithelial cells cultured from human fetal epidydimis. *J. Membr. Biol.*, **124**, 275–84.

Reid, B.L. & Cleland, K.W. (1957). The structure and function of the epididymis. I. The histology of the rat epididymis. *Aust. J. Zool.*, **5**, 223–50.

Robaire, B. & Hermo, L. (1988). Efferent ducts, epididymis, and vas deferens: structure, functions, and their regulation. In *The Physiology of Reproduction*, ed. E. Knobil & J. Neill, pp. 999–1080. New York: Raven Press.

Snouwaert, J.N., Brigman, K.K., Latour, A.M., Malouf, N.N., Boucher, R.C., Smithies, O. & Koller, B.H. (1992). An animal model for cystic fibrosis made by gene targeting. *Science*, **257**, 1083–8.

Wen, R.Q. & Wong, P.Y.D. (1988). Reserpine treatment increases viscosity of fluid in the epididymis of rats. *Biol. Reprod.*, 38, 969–74.

Wong, P.Y.D. (1988). Mechanism of adrenergic stimulation of anion secretion in cultured rat epididymal epithelium. *Am. J. Physiol.*, **254**, F121–33.

Wong, P.Y., Huang, S.J., Leung, A.Y.H., Fu, W.O., Chung, Y.W., Zhou, T.S., Yip, W.W.K. & Chan, W.K. (1992). Physiology and pathophysiology of electrolyte transport in the epididymis. In *Spermatogenesis–Fertilization–Contraception: Molecular, Cellular and Endocrine Events in Male Reproduction*, ed. E. Nieschlag & U.-F. Habenicht, pp. 319–44. Berlin: Springer.

6

The mammary gland epithelial cell

Shirley E. Pullan and Charles H. Streuli

Introduction

The mammary gland undergoes extensive proliferation, differentiation and remodelling during normal development. At puberty, epithelial ducts infiltrate the stromal tissue and create branching networks. In pregnancy, the luminal cells of these ducts change their cell fate and proliferate rapidly to form spherical alveolar structures that are able to synthesise and secrete milk products vectorially. Following a lactational phase, the secretory epithelial cells die by apoptosis and the gland reverts to a non-pregnant condition. The gland is dependent upon many factors which provide signals both for progression through these different developmental stages, and for maintaining differentiation. Hormones, such as oestrogen, progesterone, prolactin and growth hormone, are important for many of these processes (Topper & Freeman, 1980). Mammary epithelial cells additionally require local environmental cues in order to survive, proliferate and secrete milk (Streuli, 1995). The extracellular matrix, for example, provides molecular signals both for suppressing apoptosis in mammary epithelial cells (Boudreau *et al.*, 1995; Pullan *et al.*, 1996) and for inducing differentiation and transcription of tissue-specific milk protein genes (Barcellos-Hoff *et al.*, 1989; Schmidhauser *et al.*, 1990; Streuli *et al.*, 1991). Cell–cell contacts are also important for some aspects of differentiation including the establishment of polarity and directional secretion of milk products.

The cellular events that occur during puberty and pregnancy in the mammary gland therefore represent a wide spectrum of processes fundamental to cell and developmental biology. In particular, the mammary gland provides an ideal system (i) to address the mechanisms behind the formation of tubular epithelial structures and their ability to create branching networks, (ii) to examine the formation of simple multicellular alveolar structures and the establishment of polarity, (iii) to study the intracellular signalling pathways that link the extracellular matrix with gene transcription and (iv) to decipher the molecular cues that trigger apoptosis. Because mammary development mostly occurs post-natally, and because the tissue is physically amenable to experimental manipulation, it provides a valuable and multifaceted resource for study.

Many techniques are currently used to study the regulatory molecules and signalling events that control mammary phenotype; these include transgenesis, tissue reconstruction methods and whole mount organ culture. However, for the final analysis of molecular function, it is often necessary to simplify a complex multicellular system by study at the individual cell level. In the case of mammary parenchyma this necessitates isolating organoids or individual epithelial cells, and culture environments are now available that promote an '*in vivo*' cellular phenotype. Early-passage cultured mammary epithelial cells can be induced to undergo tissue morphogenesis, lactational differentiation and apoptosis, mimicking the analogous events *in vivo*. Furthermore, the molecular mechanisms by which these events take place can be characterised using appropriate culture models. For example, the association between mammary epithelial cells and the basement membrane has been shown to induce milk protein synthesis through signals that are mediated via integrins (Streuli *et al.*, 1991) and derived from the extracellular matrix component laminin (Streuli *et al.*, 1995*a*). These locally acting cues operate together with lactogenic hormones, resulting in transcriptional activation of milk protein promoters (Schmidhauser *et al.*, 1990) through pathways involving specific transcription factors (Streuli *et al.*, 1995*b*).

This chapter explains techniques for isolating and culturing mammary epithelial cells from mice, and discusses environments that promote tissue-specific function. The methods have been developed over many years, and are mainly derived from Emerman & Pitelka (1977), Lee *et al.* (1984), Emerman & Bissell (1988) and Barcellos-Hoff *et al.* (1989). They can be modified in a number of ways to answer specific mechanistic questions (Streuli *et al.*, 1991, 1995*a*; Pullan *et al.*, 1996).

Source of cells

Mammary glands are located between the skin and the peritoneum. There are five pairs of glands in the mouse, and their location is shown diagrammatically in Fig. 6.1. The morphology of the tissue and the cell composition of the gland change dramatically during the different stages of development.

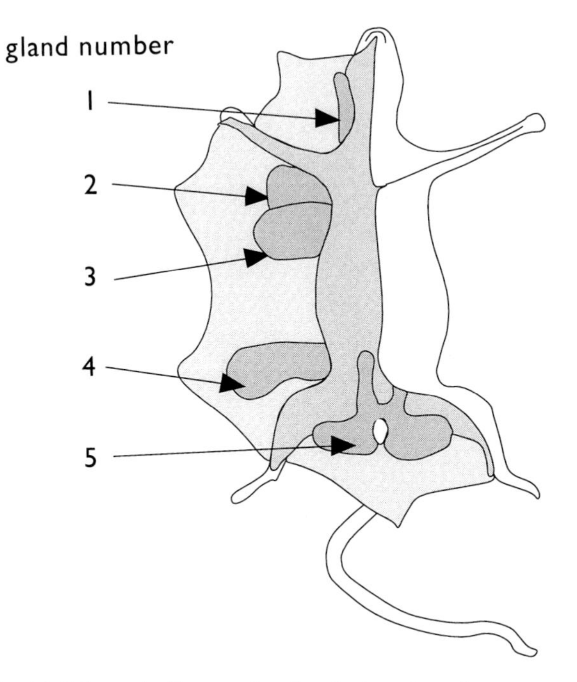

Fig. 6.1. Schematic diagram showing the location of mouse mammary glands. This diagram of a partially dissected mouse shows the approximate location of glands 1 to 5. Gland 3 lies over the top of gland 2. The mammary tissue is soft and pale pink in colour. Tendons and lymph nodes not shown.

These changes reflect functional alterations within the mammary tissue, in preparation for full lactation.

In virgin, or nulliparous animals, the mammary gland consists of a sparsely branched ductal epithelium embedded within a fatty stroma. The tissue is pale pink or cream in colour and can be hard to see. The appearance of the gland changes during pregnancy when the parenchymal component proliferates, resulting in a darker pink coloured tissue, abundant in alveolar, secretory epithelium. These mammary glands are larger, softer, and easier to remove during dissection. At the lactational stage, the tissue is whiter due to the presence of milk. During early involution, the tissue is engorged with milk, and is consequently very delicate. Subsequently, extensive apoptosis occurs throughout the gland.

In addition to the overall appearance of the tissue, the proportions of different cell types within mammary gland change during development. The ductal network in non-pregnant animals consists of polarised luminal epithelial cells completely ensheathed by a layer of myoepithelium (Fig. 6.2*a*). A

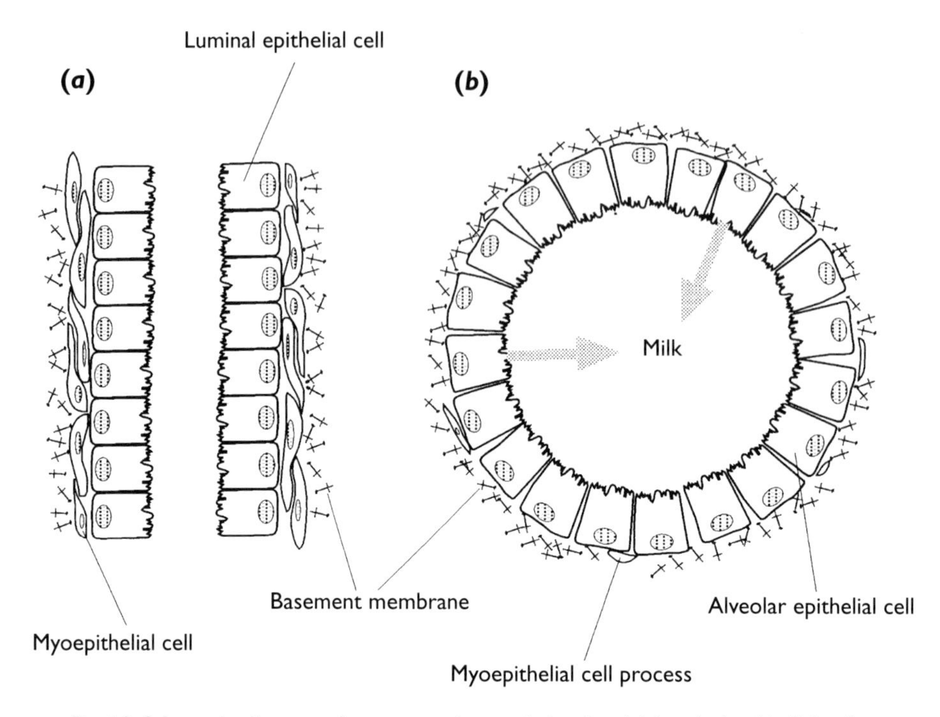

Fig. 6.2. Schematic diagram of mammary duct and alveolus. (*a*) Luminal epithelial cells line the collecting ducts of mammary gland, which are separated from the basement membrane by a layer of myoepithelium. (*b*) Alveolar epithelium consists of a single layer of polarised cells that interact directly with basement membrane and occasional processes of myoepithelial cells. The cells secrete milk products vectorially. In addition to the parenchymal component, fibroblasts and adipocytes are found in the surrounding stroma.

basement membrane separates this structure from an interstitial matrix containing fibroblasts, and distal to this is fatty stroma containing mostly adipocytes. In addition, each gland contains a lymph node and networks of capillaries and nerves. Thus, when mammary glands are dissected for tissue culture, the initial isolate contains a spectrum of cell types. Purification protocols are designed to maximise the isolation of particular cell types.

Glands from pregnant mice contain a similar variety of cell types, but in addition have a large proportion of milk-producing alveoli. These structures contain a single layer of polarised highly-secretory luminal cells which are encased in a sparse lattice of myoepithelium (Fig. 6.2*b*). Both the alveolar luminal cells and myoepithelial cells lie on a continuous basement membrane, and are arranged as clusters of many alveoli. When appropriately stained and viewed under a dissecting microscope, they resemble bunches of grapes. Alveoli become most abundant in the latter third of pregnancy, so for differentiation studies mammary tissue is best isolated from mice in mid to late pregnancy, i.e. 14.5–18.5 days of gestation. At this stage of development the gland provides the greatest yield of epithelial cells per mouse. In lactation, the cells are full of milk protein and fat droplets and are consequently more fragile and easier to rupture, so isolation of fully lactational alveoli is difficult; one of the culture models described below allows cells isolated from pregnant mice to progress, both morphologically and functionally, to a full lactational phenotype (Aggeler *et al.*, 1991).

Markers of mammary gland epithelial cells

The different cell types within mammary gland are easy to distinguish by light microscopy in sections stained with haematoxylin and eosin. For more precise cell type identification, and for cell typing in culture, specific markers are required. Cytoskeletal markers are the easiest to procure for staining cells in tissue sections, and such markers can also be used for cell type analysis in culture. However, long-term culture can produce artefactual results for some antigens, exemplified by the frequent up-regulation of vimentin expression in cultured mammary epithelial cells. Our cell isolation procedure (see below) enriches for epithelial cells, which can be detected with cytokeratin markers. Thus, basal myoepithelial cells can be detected with antibodies to cytokeratin 5 or 14, or alternatively markers for smooth muscle actin (e.g. clone IA4, Sigma A2547). Luminal cells contain cytokeratins 7, 8, 18, and sometimes 19, and these can be detected with anti-pan keratin antibodies (e.g. rabbit anti-cow muzzle keratin, Dako N 1512; or clone AE1, ICN 69-140–1). However, luminal cells from ducts and alveoli express subtle differences in keratin expression and show different survival potentials in culture, indicating that they are different cell types (our unpublished data). In addition to cytokeratin markers, distinct cell types express different cell-surface antigens. Myoepithelial cells express the CALLA antigen (common acute lymphoblastic leukaemia antigen, otherwise known as CD10 or neutral endopeptidase 24.11), whereas luminal epithelial cells express epithelial membrane antigen. These surface antigens are useful in identifying the different cell types in the mammary cell population, and antibodies directed against them have been used to purify individual cell populations by fluorescence-activated cell sorting (Dundas *et al.*, 1991) or immunomagnetic separation (Clarke *et al.*, 1994).

Methods of isolation

Here we describe a method for isolating mammary epithelial cells from non-pregnant or pregnant mice. The principle of the procedure is to dissect the mammary gland from animals, chop it into small pieces, and digest the tissue with collagenase. Washing by differential centrifugation results in purification of intact organoids from ducts and alveoli, with almost all single cells, including most fibroblasts, endothelial cells, lymphocytes, and many myoepithelial cells, being discarded.

Dissection tools

2 beakers containing 70% ethanol: 600 ml for animals, 250 ml for instruments
Forceps: 1 Adson rat tooth forceps for holding skin, 1 fine grip with fine serrations for tissue
Scissors: 1 rounded pair for cutting skin, 1 pair tissue scissors
10–15 needles or dissection pins
Pre-sterilised stick swabs (DIFCO 9349-32)
Dissection board (e.g. polystyrene box lid covered with foil)
250 ml conical flask
250 ml beaker covered with four layers of medical gauze
3 scalpel blade holders and 3 no. 23 blades
Teflon board for chopping tissue

All equipment (except the dissection board) is sterilised prior to use.

Collagenase solution

Collagenase A (Boehringer Mannheim 103 586) 3.0 mg/ml
HEPES, Na salt (Sigma H 0763) 2.6 mg/ml
$NaHCO_3$ 1.2 mg/ml
Trypsin (≈255 BAEE U/mg) (Life Technologies 840-7250) 1.5 mg/ml
Ham's F-10 powdered medium 9.8 mg/ml
Heat-inactivated fetal calf serum 5%
pH 7.4, then filter sterilise (0.45 μm)

The collagenase has an activity of 0.4–1.0 Wünsch U/mg and is batch-tested. The solution is made on the day of the digestion and kept at 4 °C until about 30 min before use. The trypsin should also be tested as some batches result in very low cell viability and yield.

Medium for cell washes
Cold Ham's F-12 medium containing 50 μg/ml gentamicin

Solutions for determining cell yield
Solution 1: 10 mM EDTA in phosphate-buffered saline (PBS), pH 7.4
Solution 2: 2 μg/ml crystal violet, 3% acetic acid, 0.2% Triton X-100, in water, then filtered (0.22 μm)

Mammary gland dissection

The dissection should take place in a flow hood under sterile conditions if cells are to be used for culture. It is recommended that animals are killed by CO_2 overdose to avoid damaging glands in the neck. Bathe the animal in 70% ethanol to sterilise it and pin out onto a dissection board. Make a V-shaped cut in the lower abdomen, through the skin only. Pull the skin away from the peritoneum and cut the skin right up to the chin. Cut diagonally outwards towards each limb. Pin out one side of the skin to begin with, stretching it taught (Fig. 6.1). Gland 1 lies in the neck and under the shoulder as a thin strip. Avoid taking the cheek muscles, which are similar in colour but muscular in appearance. In general, when dissecting, avoid lymph nodes and tendons which may contaminate the primary cell cultures. In particular, there are three tendons near glands 2 and 3, one of which is obscured and lies directly between the two glands. Gland 3 lies almost directly on top of gland 2 in the thoracic area. Use a swab to separate gland 3 from the underlying tendon and remove, then discard the tendon and dissect gland 2. Check under the shoulders for any tissue that may remain. Gland 4 can be removed most easily by pushing it away from the skin with a swab and then finally pulling or cutting it away from the skin, avoiding the lymph node at the top of the hindlimb. Gland 5 is most easily seen in late-pregnant animals. The pair of glands often lie together against the body as two thin strips when the skin is pulled away. The mammary tissue around the groin is best plucked away with forceps. Place the glands in a pre-weighed tube containing medium as they are dissected. Expect 0.65 g (±0.12 g) tissue from 8–10 week virgin ICR or CD-1 mice, and 1.58 g (±0.33 g) tissue from 14.5–18.5 day pregnant animals.

Collagenase digestion

If the mammary glands are to be collagenase digested, first chop them well, 4–5 g at a time. Tissue homogenisers are not appropriate for this procedure, and we have found that the only method that works well is to chop manually for 5–10 min on a Teflon board with three scalpel blades held or strapped together. This results in chopped glands the consistency of paste. Place in a 250 ml conical flask containing approximately 5 ml collagenase per gram of tissue, and follow the digestion/centrifugation protocol shown in Fig. 6.3. Digest the mammary gland for 1 h at 200 rpm on a rotary shaker (37 °C), then spin out large lumps of undigested material in a benchtop centrifuge (16 *g* for 1 min). Redigest for 30 min using about half the amount of collagenase as for the first digest. Meanwhile, spin the supernatant (115 *g* for 3 min) to give pellet 1. Resuspend the pellet in cold F-12 medium and keep on ice. Spin the supernatant (400 *g* for 10 min) to give pellet 2. Filter redigested tissue into a beaker covered with gauze and rinse through with F-12 medium. Spin (115 *g* for 3 min) to give pellet 3 and the resulting supernatant (400 *g* for 10 min) to give pellet 4. Combine pellets 2 and 4 and spin (115 *g* for 3 min) to give pellet 5. Finally, combine pellets 1, 3 and 5, and separate into several 50 ml tubes, using one tube for 4–6 g starting tissue weight. Wash three times (115 *g* for 3 min) with 45 ml of cold F-12 medium. One further wash is sometimes included, especially when a lot of tissue has been digested. The pellet should be fairly compact with no pieces of tissue floating away.

Determination of cell yield

The cells isolated by the collagenase digestion and differential centrifugation procedure are present mostly as organoids. The most effective way to determine the cell yield is to separate the organoids using EDTA and then count the nuclei. Suspend organoids in 10–20 ml F-12 medium, and mix 50–100 μl with an equal volume of solution 1. The cells should be triturated for 3–4 min to break up the cell aggregates. Remove 100 μl and add to 100 μl of solution 2. Remove 100 μl of this and add to 100 μl of solution 1. Count purple nuclei in triplicate using a haemocytometer. Expect a yield of 9.6×10^6 ($\pm5.3\times10^6$) cells per gram of tissue from non-pregnant mice, and 40×10^6 ($\pm16.5\times10^6$) cells per gram of tissue from pregnant animals.

Establishment and maintenance of primary cultures

Mammary epithelial organoids can be plated on any number of substrata, and some possibilities are given here. For example, cells can be plated on tissue culture dishes or glass coverslips that have been pretreated with serum or collagen I to aid in plating efficiency. Alternatively, thick gels of stromal matrix (collagen I) or basement membrane matrix (EHS tumour extract) can be

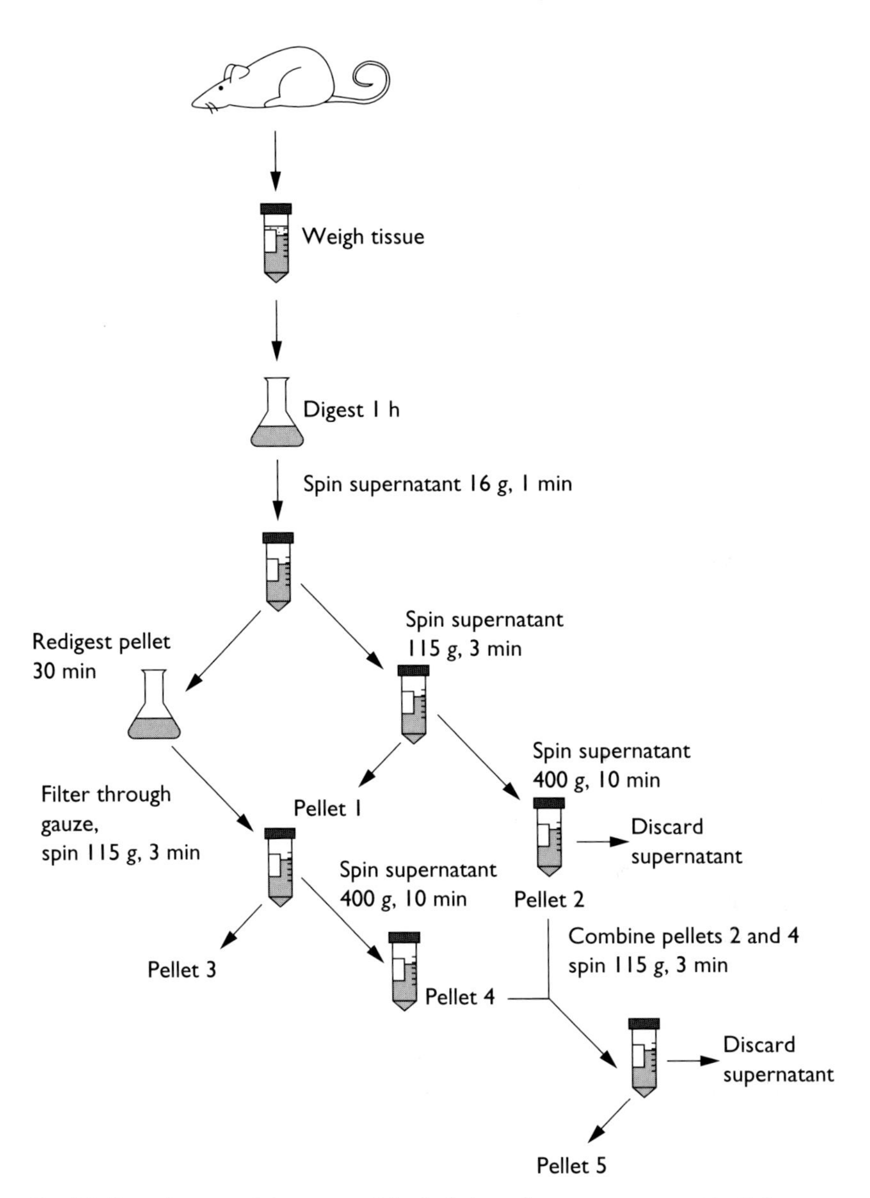

Fig. 6.3. Flow diagram of the protocol for isolation of mammary organoids. Cells are washed carefully by resuspending pellets using plastic, wide-bore pipettes and stored on ice until required for the next step. On completion, combine pellets 1, 3 and 5 in four tubes and wash organoids three times (115 *g* for 3 min).

used. For maximum plating efficiency the culture dishes are best treated with appropriate substrata well in advance. Serum and epithelial growth factor (EGF) are required for efficient cell spreading, so these are normally included for the first few days of culture.

Precoating dishes

Coating dishes with serum

Heat-inactivated fetal calf serum 20%, batch-tested for good plating efficiency
Fetuin (Sigma F 3385) 1 mg/ml
in F-12 medium

Pre-filter through a Whatman no. 1 filter, then sterilise by filtration (0.45 μm). Coat plastic dishes using 100 μl/cm^2 dish area (4–5 h, 37 °C).

Coating dishes with collagen I

Rat-tail collagen I, 1–3 mg/ml in 0.1% acetic acid (see section on substrata)
PBS, sterile

Dilute rat-tail collagen in ice-cold, sterile PBS to give a concentration of 80 μg/ml. Plate 100 μl/cm^2 dish area (overnight, 4 °C). This results in a coating density of 8 μg/cm^2. Wash dishes twice with cold PBS, once with F-12 medium, and then add serum–fetuin mix.

Coating dishes with EHS tumour matrix

EHS extract is obtained from Collaborative Biomedical Products Inc. or can be prepared in the laboratory (see section on substrata).

Chill dishes on ice and add 20 μl EHS matrix per cm^2 dish area. Spread over the whole area of the dish using the reverse end of a sterile yellow or blue Gilson tip, then leave for 2–3 min on ice to allow the matrix to spread evenly. Gel the matrix in an incubator (1 h, 37 °C), then add serum–fetuin mix.

Coating dishes with thick gels of collagen I

Rat tail collagen I in 0.1% acetic acid (see section on substrata)

10×concentrated F-12 medium, sterile
$NaHCO_3$ 0.26 M, NaOH 0.125 M, sterile

Use ice-cold solutions. Mix 8 vol collagen with 1 vol concentrated medium, then add 1 vol $NaHCO_3$/NaOH solution and mix gently. Coat chilled dishes with 150 μl collagen mix per cm^2 dish area. Gel in an incubator (1 h, 37 °C), then wash with medium (two times, 30 min) to correct salt imbalance and add serum–fetuin mix.

Media

2×Plating medium
Insulin (Sigma I 6634) 10 μg/ml
Hydrocortisone (Sigma H 0888) 2 μg/ml
EGF, receptor grade (Promega G 5011) 10 ng/ml
Gentamicin 100 μg/ml
Penicillin 200 U/ml
Streptomycin 200 μg/ml
in F-12 medium

Insulin is made as a 5 mg/ml stock in 5 mM HCl, and stored in aliquots at −20 °C. Hydrocortisone is made as a 1 mg/ml stock in ethanol, and stored at −20 °C. EGF is kept in aliquots at −70 °C. These are added to the medium just before use.

Growth medium
Insulin 5 μg/ml
Hydrocortisone 1 μg/ml
EGF, receptor grade 5 ng/ml
Heat-inactivated fetal calf serum 5%, batch tested
Gentamicin 50 μg/ml
Penicillin 100 U/ml
Streptomycin 100 μg/ml
in F-12 medium

Differentiation medium
Insulin 5 μg/ml
Hydrocortisone 1 μg/ml
Prolactin (Sigma L 6520) 3 μg/ml
Gentamicin 50 μg/ml
in DMEM/F-12 medium 1:1 (v:v)

Prolactin is made as a 3 mg/ml stock in 10 mM NaOH, sterilised by filtration, and stored for up to 2 weeks at 4 °C. It is added to the medium just before use.

Establishing primary cultures

Once a suspension of alveoli or ducts has been obtained from the mammary gland dissection/digestion procedure, the cells are resuspended in 2×plating medium and then carefully pipetted onto prepared dishes already containing 100 μl/cm^2 serum–fetuin mix. The final concentration of supplements is 10% fetal calf serum, 5 μg/ml insulin, 1 μg/ml hydrocortisone, 5 μg/ml EGF, 500 μg/ml fetuin, 50 μg/ml gentamicin, 100 U/ml penicillin, 100 μg/ml streptomycin, and the cells are maintained in a humidified incubator at 5% CO_2. A suitable plating density is 2.5–5.0×10^5 cells/cm^2, or as a rough guide, one 140 mm dish per mid-pregnant mouse. The efficiency of cell attachment and growth can be affected by the batch of serum and quality of EGF used. Additional supplements included by other investigators include 100 ng/ml cholera toxin (Clarke *et al.*, 1994), 5 μg/ml linoleic acid (Imagawa *et al.*, 1989), and 5 μg/ml prolactin, 5 μg/ml aldosterone (Tonelli & Sorof, 1982), or 1 μg/ml progesterone (Darcy *et al.*, 1995) for virgin cultures.

Maintaining cultures

After plating, the mammary organoids attach to the substrata and the cells spread out. The medium should be changed to growth medium after 2 days, and thereafter every 2 days if the cultures are to be maintained. It can take up to 5 days after the initial plating for confluent monolayers to be formed, and the cell number increases over the first few days in culture (Fig. 6.4). However, removing pregnant mammary epithelial cells from their natural environment where they contact a basement membrane, onto an artificial substratum in culture, results in apoptosis, and we have shown that these cells undergo programmed cell death in the absence of survival signals from basement membrane (Pullan *et al.*, 1996). Thus, after 4 days on plastic or collagen I the cultures deteriorate rapidly, although until the cells have entered apoptosis they remain perfectly viable.

Plating cells on substrata for differentiation studies

One of the major reasons for culturing mammary epithelia from pregnant mice is for studies on the molecular control of tissue-specific gene expres-

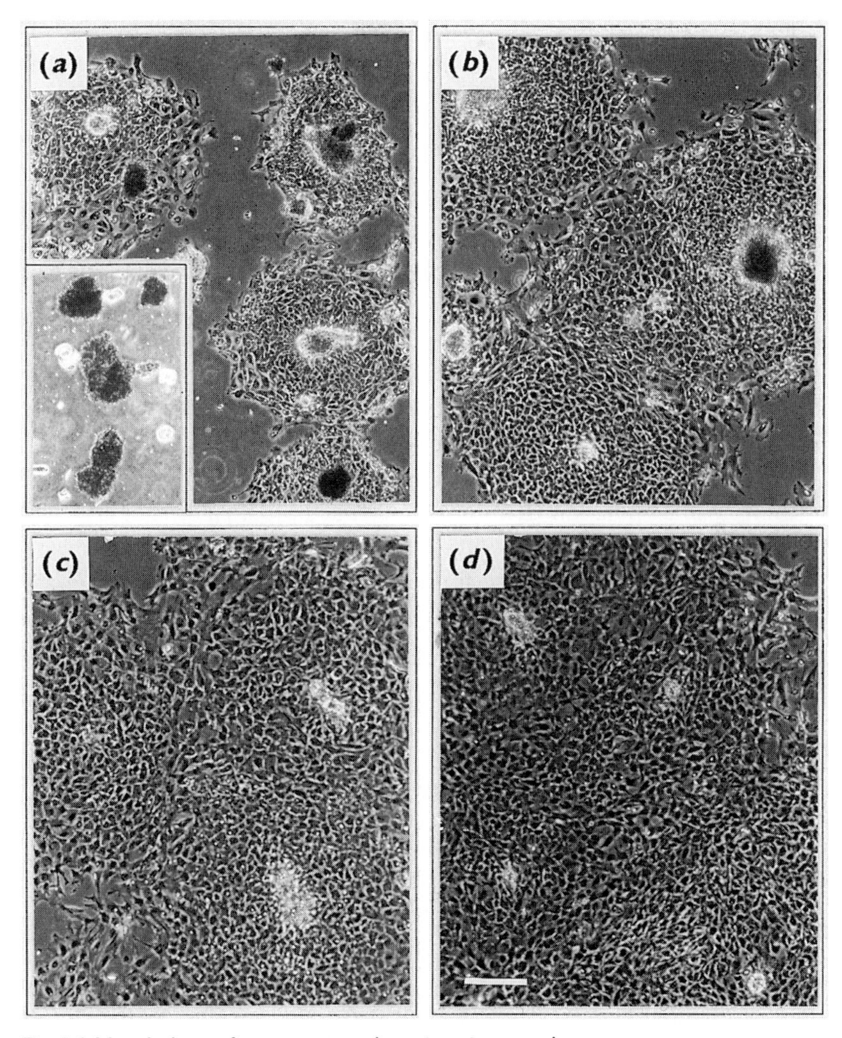

Fig. 6.4. Morphology of mammary explants in primary culture.
Mammary organoids were plated on collagen coated dishes at approximately one 140 mm dish per pregnant mouse and photographed after (*a*) 2 days, (*b*) 3 days, (*c*) 4 days and (*d*) 5 days in culture. The inset of frame (*a*) shows the morphology of organoids at the time of plating. By day 2 nearly all alveoli and ducts adhered to the dish and began to spread. At day 4 almost all cells have spread out to form a 90–95% confluent monolayer. In this experiment, identical dishes were trypsinised and the cell numbers recovered were 1.3×10^4 per cm^2 on day 2, 2.4×10^4 per cm^2 on day 3, 2.8×10^4 per cm^2 on day 4, and 3.2×10^4 per cm^2 on day 5. Scale bar represents 250 μm.

sion. Mammary cells lose their ability to differentiate in the absence of basement membrane cues, and alternative substrata to plastic and collagen have therefore been used to induce differentiation. Two examples are given here:

(1) Mammary epithelial cells undergo morphological and functional changes when they are plated onto a thick collagen gel and allowed to contract it (Emerman & Pitelka, 1977; Lee *et al.*, 1984; Streuli & Bissell, 1990). Gel contraction can be achieved by releasing the gel from the petri dish using a blue Gilson (P1000) tip or a spatula. If this is done when the cells have formed a confluent monolayer on the gel surface, 48–72 h after plating, the cells become polarised and deposit an endogenously synthesised, intact basement membrane. In differentiation medium containing lactogenic hormones but no serum or EGF, the cells secrete milk proteins. The basement membrane, induced by gel floatation, is essential for this process.

(2) An exogenous basement membrane can also be used to support mammary differentiation (Barcellos-Hoff *et al.*, 1989; Aggeler *et al.*, 1991; Streuli *et al.*, 1991). The most abundant supply of such a reconstituted basement membrane is that isolated from the EHS tumour. Cells undergo dramatic morphological rearrangements when plated on EHS matrix, forming hollow spherical structures rather than cell monolayers. In the presence of appropriate hormones, they become fully lactational alveoli.

Longer term culture and passage

For many purposes, primary cultures are fine, although they are not always homogeneous since the cells spread slowly from organoids and form monolayers only gradually. It is therefore often preferable to grow cells for 3–5 days before replating as secondary cultures. For this, primary cells are first cultured on plastic dishes or dishes coated with collagen I. In some cases, cells are plated initially on EHS matrix, but harsher conditions are required to remove them.

Solutions

HBSS/EDTA
EDTA-Na_2 2 mg/ml
in Ca^{2+}-free Hank's Balanced Salt Solution (HBSS) (Sigma H 4891)
pH 7.4

Trypsin/EDTA
Trypsin (Sigma T 0646) 0.5 mg/ml
in HBSS/EDTA

10×trypsin/EDTA
Trypsin 5.0 mg/ml
in HBSS/EDTA

Dulbecco's Modified Eagle Medium (DMEM)/F-12 medium 1:1 (v:v), either made up from constituent media DMEM and F12 or bought as mixture (Sigma, GIBCO)

DMEM/F-12+5% heat-inactivated fetal calf serum

Dispase
Dispase II (Boehringer Mannheim 165 859) 9.6 mg/ml
$MgCl_2.6H_2O$ 150 μg/ml
in Puck's saline A
pH 7.4

Trypan blue
Trypan blue (Merck 34078 4K) 1.6 mg/ml
in HBSS

All solutions are sterile filtered and warmed to 37 °C before use. The trypsin and dispase solutions can be stored at −20 °C.

Removing cells with trypsin or dispase

To remove cells from plastic or collagen-coated dishes with trypsin, wash in DMEM/F-12, and add 100 μl HBSS/EDTA per square centimetre of dish (5 min, 37 °C). By this time the cells should begin to lose cell–cell contacts, visible by phase contrast microscopy. Replace medium with 20 μl trypsin/EDTA per square centimetre of dish (2–5 min, 37 °C), then tap the dish to remove any adherent cells. Collect the cells and quench trypsin activity by adding an equal volume of DMEM/F-12+5% fetal calf serum. Wash the dish to remove remaining cells, spin (250 *g* for 4 min), and resuspend the cell pellet in DMEM/F-12+5% fetal calf serum. Count total cell number and cell viability using trypan blue. If necessary, large cell clumps can be removed by passing through a 70 μm cell strainer (Falcon 2350). Trypsinising epithelial cells requires care, since too little treatment with enzyme results in excessive cell aggregation, and too much will reduce viability.

To trypsinise cells from EHS matrix or thick collagen gels, use the same method but 10×trypsin/EDTA is required for digestion of the matrix. The time of trypsinisation will be longer (7 min, 37 °C).

Dispase for removing epithelial cells as sheets is often useful. In addition, the cell population is enriched for epithelial cells as fibroblasts are removed during initial washes. Wash cells with Puck's saline, then twice with 1/4 dilution of dispase solution (2.4 mg/ml) to remove single cells (5–10 min for each wash, 37 °C). Add 100 μl diluted dispase solution per square centimetre of dish (20 min, 37 °C), then recover cell sheets into a tube containing DMEM/F-12+5% fetal calf serum. Wash the dish to remove remaining cells, spin (250 *g* for 4 min), and resuspend cell pellet in DMEM/F-12+5% fetal calf serum.

Viability of recovered cells

These methods result in 90–95% viability, but the maximum cell recovery that can be expected from pregnant mammary gland is $3–8\times10^4$ cells/cm^2 when using the above-mentioned protocol for primary cultures. The majority of the cells are epithelial but a few fibroblasts are usually present. A typical profile is 90% luminal epithelial, 9% myoepithelial and 1% fibroblast.

Establishing first-passage cultures

Cells are plated at a density of $1.0–2.5\times10^5$ per cm^2 in growth medium. First-passage cells (secondaries) adhere and spread more quickly than primaries, and result in even monolayers. If differentiation studies are being performed, the medium can be changed 24–48 h after plating. First-passage cultures from pregnant mammary glands are susceptible to apoptosis, and the cells lose their viability fairly rapidly. This means that the cells are not suitable for longer-term culture. For this, alternative protocols need to be sought, or the use of mouse mammary cell lines. A number of cell lines (Danielson *et al.*, 1984; Reichmann *et al.*, 1989; Kittrell *et al.*, 1992) or their derivatives (Ball *et al.*, 1988; Schmidhauser *et al.*, 1990; Desprez *et al.*, 1993) have been described in the last few years that retain a capacity for expressing milk-specific gene products.

Alternative methods for isolating and culturing primary mammary epithelial cells

Method I

Finely cut tissue with scalpel blades and wash in RPMI 1640 medium buffered with 10 mM HEPES (60 min, 37 °C). Transfer to HEPES-buffered RPMI containing 2.5% fetal calf serum and 1 mg/ml collagenase blend L (Sigma) and incubate on an end-over-end rotator (1 h, 37 °C). Spin (115 *g* for 3 min), discard the supernatant and redigest the pellet and top fatty layer in fresh collagenase (1.5 h, 37 °C). Spin (400 *g* for 5 min), transfer pellet and top fatty layer to a fresh solution of 2 mg/ml collagenase, and remove lymph nodes, which sediment rapidly as small round bodies. Redigest (1 h, 37 °C), spin (400 *g* for 5 min), and discard the supernatant. Suspend the pellet in 10 ml medium and filter through 400 μm mesh to remove nerve and blood vessels, allow filtrate to settle under gravity (10 min), and resuspend the pellet in 10 ml medium and leave to settle again (10 min). Plate the remaining pellet onto culture dishes in Glasgow MEM/F-12 supplemented with 10% fetal calf serum, 5 μg/ml insulin, 5 μg/ml hydrocortisone, 10 ng/ml EGF, 5 ng/ml cholera toxin (Edwards *et al.*, 1996).

Method 2

Finely mince tissue and digest in 0.4% (w/v) collagenase type IA (Sigma), 10% fetal calf serum in L15 medium (1 h, 37 °C). Filter through a 100 μm nylon mesh, redigest the undigested tissue in 0.4% collagenase solution (2 h, 37 °C), and refilter. Centrifuge both filtrates several times to isolate organoids, and remove the stromal cells by pre-plating on plastic dishes in DMEM containing 10% fetal calf serum (2 h, 37 °C). Disaggregate purified organoids by incubating first in Ca^{2+}-free DMEM (15 min, 37 °C) then 0.1% trypsin in 0.25% EDTA (2 min, 37 °C), followed by vigorous flushing through a pasteur pipette. Add DNase to 2.5 μg/ml (5 min, 37 °C), and filter the cells through 30 μm mesh and resuspend in L15 medium containing 10% fetal calf serum. Sort the cells by flow cytometry using antibodies specific for cell type markers, allow them to settle (2–3 h, 4 °C), and plate in DMEM/F-12 containing 10% fetal calf serum, 5 μg/ml insulin, 1 μg/ml hydrocortisone and 10 ng/ml cholera toxin (Dundas *et al.*, 1991).

Method 3

Finely mince mammary gland from 50-day-old virgin rats and digest in 0.2% (w/v) collagenase type III, 0.2% (w/v) dispase grade II, 5% fetal or newborn calf serum in phenol red-free RPMI 1640 medium (11–19 h, 37 °C). Wash twice in RPMI 1640 to remove lipid and adipocytes, filter through a 530 μm mesh to remove large tissue clumps, and then through a 60 μm mesh to trap epithelial organoids. Wash organoids off the filter with phenol red-free DMEM/F-12 containing 12 mM HEPES, 5% fetal or newborn calf serum. Plate on tissue culture dishes (4 h, 37 °C) to allow attachment and removal of stromal cells, then recover the organoids. This method should yield approximately 1.5×10^7 cells/g tissue. Suspend the organoids within an EHS matrix gel and culture in phenol red-free DMEM/F-12 containing 10 μg/ml insulin, 1 μg/ml hydrocortisone, 10 ng/ml EGF, 1 μg/ml progesterone, 1 μg/ml prolactin, 5 μg/ml transferrin, 0.88 μg/ml ascorbic acid, 1 mg/ml fatty-acid-free bovine serum albumin (BSA) and 50 μg/ml gentamicin (Darcy *et al.*, 1995). In this system, cells from virgin animals can undergo extensive proliferation, branching end bud and alveolar morphogenesis, and functional differentiation equivalent to that observed in the lactating animal.

Substrata

Preparation of rat tail collagen I

Solutions and reagents
KCl 1.5 M
Acetic acid 0.1%
Double-distilled water

All solutions are sterilised

This procedure is based on the original method of Elsdale & Bard (1972). Soak rat tails in 95% ethanol (15 min) and remove skin to expose four tendon bundles. Dissect filaments with fine forceps and scissors, tease them into finer filaments, and remove as much cellular material as possible. Wash three times with 1.5 M KCl, then rinse several times in water to remove salt. Remove most of the water by centrifugation before weighing the tendon. (Expect approximately 1.5 g of tendon per rat tail.) Suspend 2.5 g tendon in 500 ml acetic acid, and stir continuously (48 h, 4 °C). Clear the collagen solution by pre-filtering through three layers of sterile surgical gauze, followed by centrifugation (27 500 *g*, 1 h, 4 °C). Store the supernatant at 4 °C or at −20 °C until needed. The concentration of solubilised collagen is estimated by hydroxyproline analysis (Woessner, 1961), or by measuring the OD_{230}, where the OD of a 1 mg/ml solution of collagen I is approximately 0.91. Alternatively, type I rat-tail collagen may be purchased from Sigma (C 7661).

Preparation of EHS matrix

Buffer no. 1
NaCl 3.4 M
TRIS-Cl 50 mM, pH 7.4

$EDTA\text{-}Na_2$ 4 mM
N-ethylmaleimide 2 mM

Buffer no. 2
NaCl 200 mM
TRIS-Cl 50 mM, pH 7.4
$EDTA\text{-}Na_2$ 4 mM
Ultrapure urea 2 M
N-ethylmaleimide 2 mM

Buffer no. 3
NaCl 150 mM
TRIS-Cl 50 mM, pH 7.4
$EDTA\text{-}Na_2$ 4 mM
N-ethylmaleimide 2 mM

The solutions are sterilised and stored at 4 °C. Fresh *N*-ethylmaleimide is added just before use (from a 200 mM stock in water).

Tumour passage

This procedure is based on the original method of Orkin *et al.* (1977) for tumour passage, and Kleinman *et al.* (1986) for preparation and use of the extract. Inject 12-week-old MF1 mice intramuscularly into the right hindleg with 100 μl of a thick suspension of EHS tumour cells homogenised in DMEM. For studies on mammary gland, it is better to use male mice. Provide 0.1% β-aminopropionitrile fumarate, administered in the water, until harvesting 2–3 weeks later when tumour size has reached 2 cm. Snap-freeze tumour in liquid nitrogen and store until use, at −70 °C.

Preparation of EHS matrix from the tumour

Weigh approximately 30 g of tumour per preparation, but note the exact weight. Partially thaw on ice, then homogenise on ice using approximately 60 ml buffer no. 1. Spin (113 000 *g* for 30 min, 4 °C), discard the supernatant, and repeat three more times until a straw-coloured pellet is obtained. Homogenise in 1.8 ml buffer no. 2 per gram of original tumour and stir continuously overnight at 4 °C. Spin the homogenate (113 000 *g* for 30 min, 4 °C), and divide the supernatant between three Spectropor no. 2 dialysis bags. Dialyse three times against 1 litre buffer no. 3 (24 h total time, 4 °C), and a further two times against 1 litre DMEM/F-12 containing gentamicin and penicillin–streptomycin (12–24 h, 4 °C). Store in aliquots at −70 °C, and thaw slowly on ice when required for use as a culture substratum. The protein concentration is approximately 14 mg/ml when measured against a laminin standard in a Bradford protein assay. The quality of the EHS matrix can be assessed by checking for lack of protein degradation of 5% reducing SDS polyacrylamide gel electrophoresis. To prepare factor-reduced EHS matrix (Taub *et al.*, 1990), precipitate freshly made EHS matrix dialysed against buffer no. 3 twice with 20% $(NH_4)_2SO_4$ on ice, resuspend the matrix in buffer no. 3 and dialyse (24 h, 4 °C).

Cryopreservation

Primary cultures of mammary epithelial cells can be effectively stored in liquid nitrogen. The initial cell preparation from tissue dissection mostly exists as large cell aggregates, and has been found to freeze very poorly. However, single-cell suspensions freeze well, and can be recovered from liquid nitrogen with 60–65% efficiency.

Solutions

Freeze mix
3 parts DMEM/F-12
1 part dimethylsulphoxide (DMSO)
1 part fetal calf serum

DMEM/F-12+5% fetal calf serum

Storing primary cultures

Trypsinise cultures to give a single cell suspension, as above, and adjust to 5–10×10^6 cells per millilitre in ice-cold DMEM/F-12+5% fetal calf serum. Add an equal volume of cold freeze mix to the cell suspension, and mix gently. On ice, aliquot 1 ml cells per cryovial and freeze at a rate of −1 °C/min in an isopropanol chamber (1.5–2 h, −70 °C). Transfer vials to liquid nitrogen.

Cell recovery

For cell recovery, quickly thaw cryovials at 37 °C and transfer contents to DMEM/F-12+5% fetal calf serum. Gently triturate the cells and spin (250 *g* for 4 min). Estimate cell viability using trypan blue and plate on appropriate culture dishes in growth medium.

Troubleshooting

The following section is an aid to overcoming problems that may be experienced. The correct balance of chopping and digestion is essential to the protocol and therein lies the greatest difficulties.

Isolation of mammary epithelial cells

Large lumps appear to float up during final washes

(1) The tissue has not been either chopped enough or digested sufficiently. Chop tissue for at least 10 min per batch. This may take longer if the experimenter is not experienced. Check the batch of collagenase has a good activity. You may need to use a larger volume of the collagenase mix or use it at a higher concentration if the problem is not resolved.

(2) Without the correct balance of digestive enzymes, the cells will be either underdigested or overdigested. Either way the appearance is much the same, except that overdigested tissue may be completely destroyed. Check that activity and the concentration of the trypsin used are correct.

Establishment of primary cultures

Not all organoids plate out

A percentage of organoids will fail to plate out even under the best conditions, but the plating efficiency may be optimised. There is often great variability amongst different sera, so it is worth batch-testing sera to determine which one gives the greatest plating efficiency. Coating dishes with collagen I greatly enhances the plating efficiency compared with uncoated dishes. In addition, a further benefit is provided by coating dishes with fetuin, of which the concentration may be increased to 1 mg/ml (final) to increase plating efficiency. The use of EGF further enhances the plating efficiency and highly purified EGF (receptor grade) is recommended. Try using EGF at 10 ng/ml if this increases the plating efficiency without being cost-inhibitive. If only a very small percentage of organoids plate out then this is a different problem and may be due to a low tissue viability, e.g. caused by overdigestion by trypsin.

Organoids take a long time to spread out

If the organoids are very large, this may result from an imbalance in the digestive enzymes. Check the concentration and activity of trypsin and collagenase used in the collagenase solution. The correct balance between the two enzymes is essential to digest the tissue fully. Alternatively, the tissue may not have been chopped well enough, and without sufficient chopping the enzymes cannot penetrate and digest tissue in order to free the organoids.

Striated objects appear when plated out

These objects could be either tendon or muscle. Take care during the dissection to remove only mammary tissue, especially around the cheek area for gland 1 and under the shoulder and forelimb with glands 2 and 3.

Longer-term passage

On trypsinisation many cells are left behind

Pretreating with EDTA (0.02%) in HBSS before trypsinising is the most efficient way of removing all the cells whilst maintaining high cell viability. Make sure cells are beginning to lose cell–cell contacts before adding trypsin–EDTA solution. Be patient but do not over-trypsinise. Tap the dishes if necessary to dislodge any remaining cells and wash dishes several times to remove cells. Over- or under-trypsinisation may result in cells clumping together. Large cell clumps may be removed by filtration through a 70 μm cell strainer.

Many organoids are crowded onto one area of the dish

Before plating, gently shake the dish to obtain an even distribution of organoids and be sure to place in the incubator carefully.

Cell yield is poor on trypsinisaton

Pretreating with EDTA before trypsinising is the most efficient method of passaging these cells, but the time of passage is also important. After 4 or 5 days the cultures should have spread, fully, or almost fully, and trypsinisation at this time yields the greatest cell number. Mammary cells in culture will die by apoptosis continually, so there is no advantage in leaving cultures much later than this before trypsinising.

Acknowledgements

We are indebted to Dr Teresa Klinowska for critical review of the manuscript. C.H.S. is a Wellcome Senior Research Fellow in Basic Biomedical Science.

References

Aggeler, J., Ward, J., Blackie, L.M., Barcellos-Hoff, M.H., Streuli, C.H. & Bissell, M.J. (1991). Cytodifferentiation of mouse mammary epithelial cells cultured on a reconstituted basement membrane reveals striking similarities to development *in vivo*. *J. Cell Sci.*, **99**, 407–17.

Ball, R.K., Friis, R.R, Schoenenberger, C.A., Doppler, W. & Groner, B. (1988). Prolactin regulation of β-casein gene expression and of a cytosolic 120-kd protein in a cloned mouse mammary epithelial cells line. *EMBO J.*, **7**, 2089–95.

Barcellos-Hoff, M.H., Aggeler, J., Ram, T.G. & Bissell, M.J. (1989). Functional differentiation and alveolar morphogenesis of primary mammary cultures on reconstituted basement membrane. *Development*, **105**, 223–35.

Boudreau, N., Sympson, C.J., Werb, Z. & Bissell, M.J. (1995). Suppression of ICE and apoptosis in mammary epithelial cells by extracellular matrix. *Science*, **267**, 891–3.

Clarke, C., Titley, J., Davies, S. & O'Hare, M.J. (1994). An immunogenic separation method using supramagnetic (MACS) beads for large-scale purification of human mammary luminal and myo-epithelial cells. *Epithelial Cell Biol.*, **3**, 38–46.

Danielson, K.G., Oborn, C.J., Durban, E.M., Butel, J.S. & Medina, D. (1984). Epithelial mouse mammary cell line exhibiting normal morphogenesis *in vivo* and functional differentiation *in vitro*. *Proc. Natl. Acad. Sci. USA*, **81**, 3756–60.

Darcy, K.M., Shoemaker, P.-P.H., Lee, M.M., Vaughan, J.D., Black, J.D. & Ip, M.M. (1995). Prolactin and EGF regulation of the proliferation, morphogenesis, and functional differentiation of normal rat mammary epithelial cells in three dimensional culture. *J. Cell. Physiol.*, **163**, 346–64.

Desprez, P.-Y., Roskelley, C., Campisi, J. & Bissell, M.J. (1993). Isolation of functional cell lines from a mouse mammary epithelial cell strain: the importance of basement membrane and cell–cell interaction. *Mol. Cell. Differ.*, **1**, 99–110.

Dundas, S.R., Ormerod, M.G., Gusterson, B.A. & O'Hare, M.J. (1991). Characterization of luminal and basal cells flow-sorted from the adult rat mammary parenchyma. *J. Cell Sci.*, **100**, 459–71.

Edwards, P.A.W., Abram, C.L. & Bradbury, J.M. (1996). Genetic manipulation of mammary epithelium by transplantation. *J. Mammary Gland Biol. Neoplasia*, **1**, 75–89.

Elsdale, T. & Bard, J. (1972). Collagen substrata for studies on cell behaviour. *J. Cell Biol.*, **54**, 626–37.

Emerman, J.T. & Bissell, M.J. (1988). Cultures of mammary epithelial cells: extracellular matrix and functional differentiation. *Adv. Cell Culture*, **6**, 137–59.

Emerman, J.T. & Pitelka, D.R. (1977). Maintenance and induction of morphological differentiation in dissociated mammary epithelium on floating collagen membranes. *In Vitro*, **13**, 316–28.

Hollman, K.H. (1974). *Cytology and Fine Structure of the Mammary Gland*, vol. 1, *Lactation: A Comprehensive Treatise*, ed. B.L. Larson & V.R. Smith. New York: Academic Press.

Imagawa, W., Bandyopadhyay, G.K., Wallace, D. and Nandi, S. (1989). Phospholipids containing polyunsaturated fatty acyl groups are mitogenic for normal mouse mammary epithelial cells in serum-free primary cell culture. *Proc. Natl. Acad. Sci. USA*, **86**, 4122–6.

Kittrell, F.S., Oborn, C.J. & Medina, D. (1992). Development of mammary preneoplasia *in vivo* from mouse mammary epithelial cell lines *in vitro*. *Cancer Res.*, **52**, 1924–32.

Kleinman, H.K., McGarvey, M.L., Hassell, J.R., Star, V.L., Cannon, F.B., Laurie, G.W., & Martin, G.R. (1986). Basement membrane complexes with biological activity. *Biochemistry*, **25**, 312–18.

Lee, E.Y.-H., Parry, G. & Bissell, M.J. (1984). Modulation of secreted proteins of mouse mammary epithelial cells by the collagenous substrata. *J. Cell Biol.*, **98**, 146–55.

Orkin, R.W., Gehron, P., McGoodwin, E.B., Martin, G.R., Valentine, T. & Swarm, R. (1977). A murine tumor producing a matrix of basement membrane. *J. Exp. Med.*, **145**, 204–20.

Pullan, S., Wilson, J., Metcalfe, A., Edwards, G., Goberdhan, N., Tilly, J., Hickman, J.A., Dive, C. & Streuli, C.H. (1996). Requirement of basement membrane for the suppression of programmed cell death in mammary epithelium. *J. Cell Sci.*, **109**, 631–42.

Reichmann, E., Ball, R., Groner, B. & Friis, R.R. (1989). New mammary epithelial and fibroblastic clones in co-culture form structures competent to differentiate functionally. *J. Cell Biol.*, **108**, 1127–38.

Schmidhauser, C., Bissell, M.J., Myers, C.A. & Casperson, G.F. (1990). Extracellular matrix and hormones transcriptionally regulate bovine beta-casein 5′ sequences in stably transfected mouse mammary cells. *Proc. Natl. Acad. Sci. USA*, **87**, 9118–22.

Streuli, C.H. (1995). Basement membrane as a differentiation and survival factor. In *The Laminins*, ed. P. Ekblom, pp. 217–33. Switzerland: Harwood.

Streuli, C.H. & Bissell, M.J. (1990). Expression of extracellular matrix components is regulated by substratum. *J. Cell Biol.*, **110**, 1405–15.

Streuli, C.H., Bailey, N. & Bissell, M.J. (1991). Control of mammary epithelial differentiation: the separate roles of cell–substratum and cell–cell interaction. *J. Cell Biol.*, **115**, 1383–95.

Streuli, C.H., Schmidhauser, C., Bailey, N., Yurchenco, P., Skubitz, A.P.N., Roskelley, C. & Bissell, M.J. (1995*a*). Laminin mediates tissue-specific gene expression in mammary epithelia. *J. Cell Biol.*, **129**, 591–603.

Streuli, C.H., Edwards, G.M., Delcommenne, M., Burdon, T.G., Schindler, C. & Watson, C.J. (1995*b*). Stat5 as a target for regulation by extracellular matrix. *J. Biol. Chem.*, **270**, 21639–44.

Taub, M., Wang, Y., Szczesny, T.M. & Kleinman, H.K. (1990). Epidermal growth factor or transforming growth factor-beta is required for kidney tubulogenesis in Matrigel cultures in serum-free medium. *Proc. Natl. Acad. Sci. USA*, **87**, 4002–6.

Tonelli, Q.J. & Sorof, S. (1982). Induction of biochemical differentiation in three-dimensional collagen cultures of mammary epithelial cells from virgin mice. *Differentiation*, **22**, 195–200.

Topper, Y.J. & Freeman, C.S. (1980). Multiple hormone interactions in the developmental biology of the mammary gland. *Physiol. Rev.*, **60**, 1049–106.

Woessner, J.F. (1961). The determination of hydroxyproline in tissue and protein samples containing small proportions of imino acid. *Arch. Biochem. Biophys.*, **93**, 440–7.

7

The epidermal keratinocyte

Michele De Luca, Graziella Pellegrini and Giovanna Zambruno

Introduction

Epidermis, the outermost layer of skin, consists of stratified squamous epithelium, approximately 100 μm thick, accounting for only one-fortieth of skin thickness. Yet its function is essential in maintaining the stability of the interior milieu of the entire organism and in protecting the body against environmental hazards. Epidermis, which is probably the most complex stratified epithelium, is composed mainly of keratinocytes. Other cell types, such as melanocytes, Langerhans cells and Merkel cells, which have very important functions, are found in the epidermis but do not directly participate in the structural organisation of this tissue. Keratinocytes at different stages of differentiation form histologically distinct cellular layers (Fig. 7.1). The basal layer is composed of small keratinocytes with a large nucleus and scarce cytoplasm which maintain proliferative capacity throughout the lifespan of the organism. Committed basal cells lose their proliferative capacity, migrate upwards and initiate the process of terminal differentiation which leads to the sequential formation of the stratum spinosum, stratum granulosum and stratum corneum, the last being composed of numerous layers of tightly packed and flattened cells crucial for epidermal barrier function.

Keratinocyte adhesion

Basal epidermal keratinocytes rest on a basement membrane composed of a specific subset of extracellular matrix proteins such as laminin 1, laminin 5 (also known as kalinin or epiligrin), laminin 6, type IV collagen, nidogen and heparan sulphate proteoglycan. The firm adhesion of basal keratinocytes, hence of the whole epidermis, to the basal lamina is mediated by hemidesmosomes. These structures link the epithelial intermediate filament

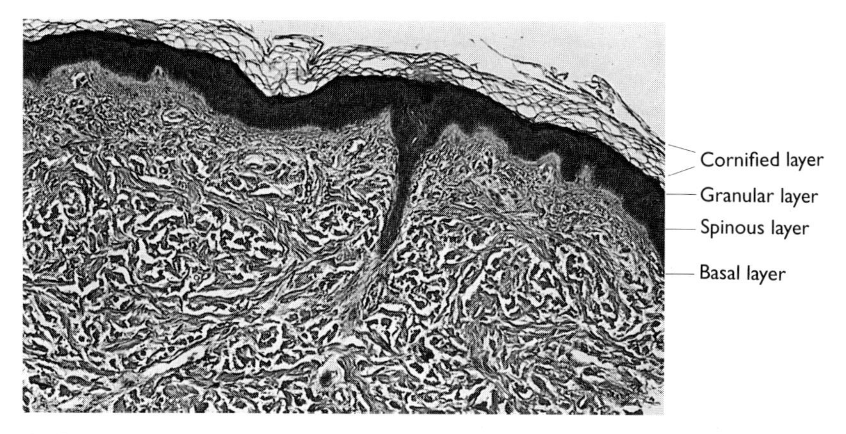

Fig. 7.1. Normal epidermal cell layers *in vivo*.

network to the dermal anchoring fibrils, which are mainly composed of type VII collagen and extend from the basement membrane to anchoring plaques in the papillary dermis. These processes are also mediated by integrin receptors, a class of transmembrane non-covalently associated glycoprotein heterodimers composed of α and β chains. Integrins are receptors for several extracellular matrix proteins, mediate cell–cell interactions in specific cell types and the adhesion, spreading and migration of cells on several components of the extracellular matrix. Normal human basal keratinocytes express α6β4, α2β1, α3β1 and αvβ5 integrin receptors, exposed on discrete plasma membrane regions in a polarised fashion.

The α6β4 integrin heterodimer is localised on the basal aspect of the basal cell, is a component of hemidesmosomes, and mediates keratinocyte adhesion to the basement membrane by binding to both laminin 1 and 5. The essential role of α6β4 in anchoring the epidermis to the basal lamina has been confirmed by mutations in the β4 gene alleles causing junctional epidermolysis bullosa associated with pyloric atresia. The α2β1 and α3β1 integrins, by contrast, are enriched laterally, at cell–cell boundaries, and cooperate with other molecules (such as cadherins and desmogleins) in regulating cell–cell interactions, possibly by forming homotypic or hybrid bonds. During the healing of acute wounds, in hyperproliferative skin diseases under the influence of transforming growth factor beta (TGFβ), or in high passage keratinocytes, integrin polarisation is lost and keratinocytes express high levels of the α5β1 fibronectin receptor and modulate their β1 and β5 integrins. In particular, $TGF\beta_1$ induces the *de novo* expression of the αvβ6 fibronectin receptor. In normal resting adult epidermis integrins are expressed only in the basal layer. The onset of terminal differentiation is therefore associated with inhibition of integrin function and loss of integrin expression from the cell surface.

Is integrin loss the cause of keratinocyte terminal differentiation, or is it its direct consequence? Some reports suggest that the onset of keratinocyte terminal differentiation is regulated by the expression and ligand occupancy of $\beta1$ integrin receptors. However, the following observations strongly suggest that the loss of integrin expression does not trigger the onset of terminal differentiation: (i) the simultaneous expression of integrins and terminal differentiation markers by adherent cells cultivated in low Ca^{2+} conditions or by paraclones (see below); (ii) the selective upward migration of involucrin-positive cells immediately after Ca^{2+} addition; (iii) the subsequent disappearance of integrins from these cells; (iv) the presence of integrins in epidermal suprabasal layers in hyperproliferative situations. Rather, integrin loss from basal keratinocytes seems to be a direct consequence of terminal differentiation and allows the selective upward migration of already committed cells. It is worth noting, however, that these studies have been performed on $\beta1$ integrins, which mainly mediate cell–cell interactions in basal keratinocytes ($\alpha2/\alpha3\beta1$) or act as 'emergency receptors' in wound healing ($\alpha5\beta1$) and not on the $\alpha6\beta4$ heterodimer which is the integrin actually mediating basal keratinocyte adhesion to the basal lamina. In this respect, it will be interesting to investigate whether the expression and function of $\alpha6\beta4$ are related to the onset of terminal differentiation.

Epidermal stem cells

The epidermis survives through a self-renewal process. Clonogenic basal keratinocytes forming the innermost epidermal basal layer regularly undergo mitosis, differentiation and upward migration to replace terminally differentiated cornified cells that are continuously shed into the environment. As a result, human epidermis is completely renewed approximately every month. To accomplish this process and to face the emergency of wound healing, the epidermis relies on the presence of stem cells and transient amplifying cells which are located both in the epidermis and in the hair matrix. The basic and essential characteristic of a stem cell is its capacity for extensive self-maintenance with proliferative self-renewal potential extending for at least one life-span of the organism, irrespective of whether the cell is multipotent (as for the haematopoietic stem cell) or unipotent (as for epidermal stem cells) in nature. Conversely, the transient amplifying cell population, which arises from stem cells and will eventually generate terminally differentiated cells, has a high proliferative rate for only a limited period of time and represents the largest group of dividing cells. Basal, clonogenic keratinocytes express integrin receptors (see above) and a newly described putative transcription factor, basonuclin. Basonuclin is expressed exclusively by keratinocytes of stratified squamous epithelia. Both *in vivo* and in stratified epidermal cultures, basonuclin is present in the nuclei of basal keratinocytes. In hair follicles, basonuclin is found in the outer root sheath and in the hair matrix, in cells possessing potential for multiplication. The disappearance of basonuclin is associated with loss of colony forming ability and precedes the onset of terminal differentiation. Although present in all multiplying keratinocytes, basonuclin is not a cell cycle marker, and its presence in the central region of large colonies (see below) strongly suggests that basonuclin-positive, growth arrested keratinocytes may revert to a growing state and generate colonies. Basonuclin, similarly to integrin receptors, identifies basal clonogenic keratinocytes located in the pathway between stem cells and terminally differentiated keratinocytes.

Keratinocyte multiplication can be studied at a clonal level using the cultivation method originally described by Rheinwald & Green (1975) (Fig. 7.2). Single basal keratinocytes can initiate colonies, allowing the precise evaluation of the founding cell growth potential (Fig. 7.2). The clonogenic potential of keratinocytes is related to the size of founding cell, while the growth potential of the single keratinocyte is determined by the type of clone which is generated. The size and shape of the colony, its perimeter and the morphology of cells forming the colony reflect the growth potential of the founding cell. Using these criteria, it is possible to identify, under a light microscope, the aborted colonies, namely those colonies which have been generated by cells with very limited growth potential. Their number is related to the age of the culture, the expected life-span of the culture, and even the age of the donor, since the number of aborted colonies is very low in cultures generated from fetal epidermis or very young donors and greatly increases with the age of the donor.

Using this clonal analysis three clonal types of keratinocytes with different capacities for multiplication have been identified and isolated in human epidermis, i.e. holoclones, meroclones and paraclones.

Holoclones form large colonies (after 14 days of culture each colony has a diameter of approximately 1 cm and contains $1\text{–}2 \times 10^5$ cells) with a very smooth and regular perimeter formed by migrating involucrin-negative cells. When subcultivated, holoclones generate large and smooth daughter colonies with a very high frequency and a negligible number of aborted colonies (0–5%). The holoclone has an extensive growth potential with proliferative self-renewal capacity. Indeed, it has been calculated that a single

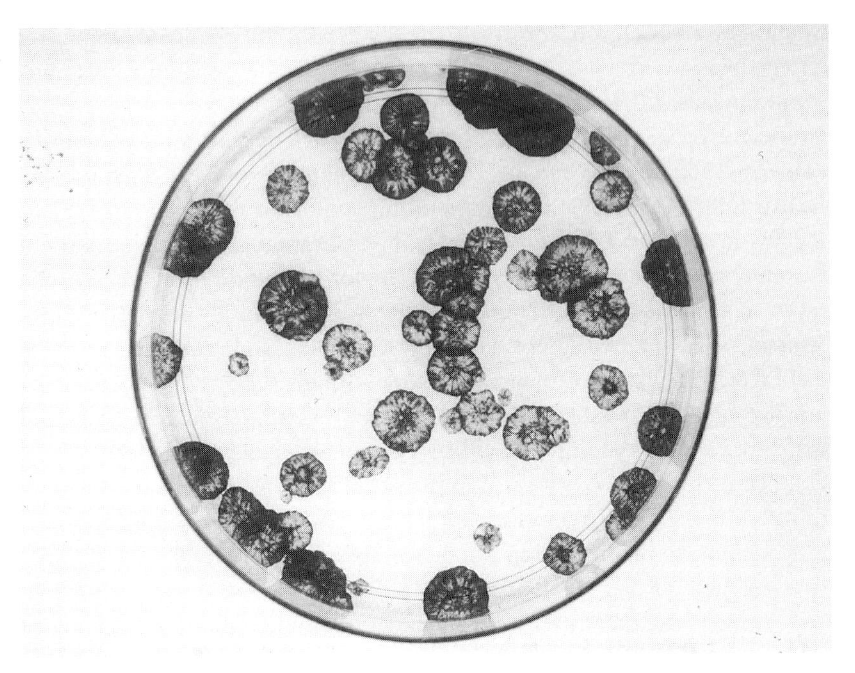

Fig. 7.2. Keratinocytes were isolated from a skin biopsy and cultivated on lethally irradiated 3T3 cells. Subconfluent primary cultures were aliquoted and cryopreserved. One hundred keratinocytes from the cryopreserved subconfluent primary culture were plated on the feeder layer and colonies stained 14 days later. Each colony is the progeny of a single keratinocyte.

holoclone can generate as many as 1.7×10^{38} progeny (over 120 doublings), i.e. enough epithelium to cover an adult human body (1.5–2 m^2), whose epidermis contains approximately 8×10^{10} keratinocytes, several times. Therefore, the basal keratinocyte generating a holoclone has the essential characteristic required to be considered as a stem cell, namely a tremendous proliferative potential.

The *paraclone*, which is generated by a transient amplifying cell, has a very limited growth potential, being committed to 1–15 divisions. The colony founded by a paraclone is very small and highly irregular, the cells forming it are large, squame-like and all involucrin-positive, and it consists of a maximum of 3×10^4 cells covering an area of no more than 17 mm^2 (these colonies are often visible only under a light microscope). When sub-cultivated, paraclones do not generate growing colonies, further indicating that paraclones consist of terminally differentiated cells.

The *meroclone* is an intermediate type of cell. The colony generated by a meroclone has a wrinkled and irregular perimeter and its size is always smaller than the size of a holoclone-derived colony, suggesting a high heterogeneity within the colony. Indeed, when subcultivated, meroclones generate a high number of paraclones, and in some instances as few as two passages are enough to achieve complete conversion of meroclones to paraclones. Therefore, the meroclone represents a reservoir of transient amplifying cells.

The transition from holoclone to meroclone to paraclone is a unidirectional process which occurs during natural ageing as well as during repeated keratinocyte subcultivation. Indeed, the culture lifetime of keratinocytes declines with the donor's age as well as with serial passaging, and this phenomenon is due to the progressive clonal conversion both *in vivo* and *in vitro*.

Therefore, the aborted colonies which can be identified as paraclones, are actually the expression of keratinocyte senescence and exhaustion of their growth potential.

In vitro reconstitution of epidermal sheets

Normal human keratinocytes isolated from a skin biopsy and plated on lethally irradiated 3T3 cells can be serially cultivated. Basal keratinocytes form colonies, each colony being the progeny of a single keratinocyte (Fig. 7.2). Once a colony is established, cells at the periphery of the colony migrate outwards and grow, and the colony expands radially up to 2 mm/day under the influence of epidermal growth factor (EGF) (Fig. 7.1). The regulation of normal human keratinocyte growth is a complex phenomenon involving both nutritive support from dermal blood vessels and a network of paracrine and autocrine loops which act mainly within the local skin environment. Fibroblasts are a source of several polypeptides binding to receptors endowed with tyrosine kinase activity and hence able to stimulate keratinocyte functions. Indeed, optimal keratinocyte clonal growth *in vitro* is obtained when cells are plated on a feeder layer of lethally irradiated fibroblasts in the presence of factors which cooperate in sustaining their growth. When cultured keratinocytes reach a critical density, they become capable of autonomous growth. These autocrine–paracrine loops have been shown to involve the synthesis and secretion of transforming growth factor alpha (TGFα), basic fibroblast growth factor (bFGF) and amphiregulin.

Melanocytes are physiologically organised in the basal layer of the cultured epidermis. They maintain the original donor site melanocyte/keratinocyte ratio, synthesise melanin granules and transfer them into the cytoplasm of surrounding keratinocytes. Keratinocytes potently induce melanocyte proliferation, direct the proper spatial organisation of melanocytes in the epidermal basal layer and regulate the proper melanocyte morphology, dendrite arborisation and melanin synthesis, through several paracrine loops and cell–cell interactions. These interactions do not require the presence of dermis and can be reproduced by cloned keratinocytes and melanocytes. Basal keratinocytes synthesise and release nerve growth factor (NGF), which regulates migration and dendritic arborisation of surrounding melanocytes and stimulates keratinocyte growth.

The differentiation and stratification process starts from the centre of the growing keratinocyte colonies. Keratinocytes committed to terminal differentiation selectively leave the basal layer, increase in size, migrate upward and show an increased density of desmosomes and tonofilaments. Keratinocyte colonies will eventually fuse, giving rise to a stratified squamous epithelium (Fig. 7.3) maintaining biochemical, morphological and functional characteristics of its *in vivo* counterpart.

The primary culture can be passaged into secondary cultures generating a large amount of epithelium. The initial cell population can be amplified up to 10 000 times in 2–3 weeks, so that, starting from a 2 cm^2 biopsy, enough epithelium can be generated to cover the entire body surface.

Autologous cultured epidermis (cultured autografts) can be successfully and permanently transplanted onto patients presenting large skin defects such as full-thickness burn wounds covering up to 98% of the body surface. Recently, the possibility has been reported of grafting cultured autografts onto patients previously transplanted with cryopreserved dermis obtained from cadaver donors. This procedure has brought about a substantial increase in the average 'take' of cultured autografts. For most of these patients the transplantation of cultured autografts is a life-saving procedure. The histological examination of the epidermal grafts reveals stratification and differentiation of suprabasal cells. Long-term studies of skin regenerated from cultured autografts revealed normal histological features. Interestingly, the transplanted cultured autografts retained characteristics of the original donor site, suggesting that human keratinocytes possess an intrinsic site-specific differentiation programme. These data have been confirmed by culturing and grafting epithelial sheets derived from oral and urethral mucosa and corneal epithelium.

The normal organisation of the epidermis is achieved even if the grafted epidermal sheet is derived from a single clonogenic keratinocyte, suggesting that epidermal reconstitution does not depend upon interactions of different keratinocyte subtypes. Rather, the maintenance of stem cells during the cultivation process is of crucial importance to ensuring the persistence of the grafted epidermis during the patient's life-time. Thus, the percentage of aborted colonies in culture, the evaluation of the growth behaviour of large

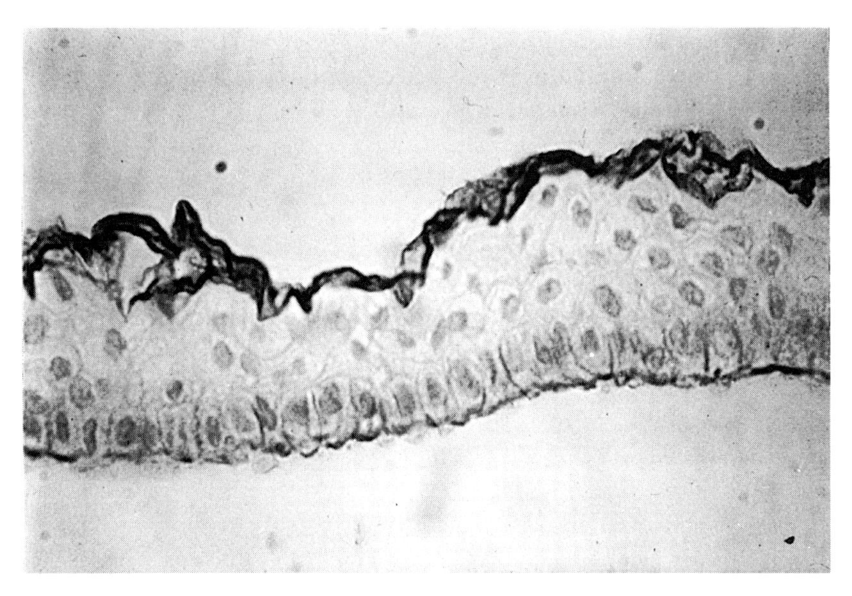

Fig. 7.3. A confluent secondary culture was detached from the surface vessel with dispase II and stained. Note the well-organised basal layer, the suprabasal differentiated layers and the absence of stratum corneum. The stratum corneum will develop after transplantation.

and smooth colonies (as doubling time and/or number of cells per square millimetre), and the periodic clonal analysis of a reference strain of keratinocytes, in terms of both clonogenic and growth potential, is of crucial importance for carefully monitoring the quality of a keratinocyte culture system. Indeed, inappropriate culture conditions can accelerate the clonal conversion irreversibly and cause the rapid disappearance of stem cells.

Human epidermal keratinocyte culture techniques

The establishment of cultures from disaggregated epidermal cells represents a new approach for growing keratinocytes compared with the skin organ and explant culture systems employed during the first half of this century. Keratinocytes plated at high density can proliferate and even reconstitute a stratified epithelium. However, clonal growth and a high number of cell doublings (more than 120) can be achieved only when keratinocytes are seeded and maintained in the presence of a lethally irradiated feeder layer of fibroblasts, as originally reported by Rheinwald & Green in 1975. In our opinion, this method is the most reliable and characterised and can be widely used for autologous transplantation onto patients.

As for many other cell types, defined media have also been developed for growing keratinocytes in serum-free conditions and in the absence of a feeder layer of fibroblasts. Defined media are commercially available and widely used for research purposes, such as testing the effects of growth factors. However, optimal keratinocyte proliferation is achieved in the presence of low Ca^{2+} concentrations (0.1–0.3 mM) at which the cells form a monolayer but do not differentiate and stratify.

In order to obtain an epidermis *in vitro* that is structurally and morphologically comparable to human epidermis *in vivo*, several organotypic culture models have been developed in which keratinocytes are grown on different 'dermal' substrates and lifted at the air–medium interface. These substrates are constituted either by human de-epidermised dermis or by dermal fibroblasts incorporated into collagen gels/films. Major applications of organotypic cultures are for toxicological studies, for studies of epithelial–mesenchymal interactions and for studies of pathological skin conditions requiring full epidermal differentiation, e.g. the replication of human papilloma viruses. The main limitations of these systems are their complexity and the small number of keratinocytes which can be grown.

Cultivation of keratinocytes on feeder layers

Preparation of feeder layers

Gamma-irradiated post-mitotic fibroblasts suppress contamination of the culture by human viable fibroblasts; further, these cells secrete soluble factors, such as insulin-like growth factor I (IGF-1), human growth factor (HGF) or keratinocyte growth factor (KGF), which sustain keratinocyte proliferation. Not all fibroblast cell lines are equally efficient at supporting keratinocyte growth: 3T3-J2 are among the most active, but NIH 3T3, Balb/C 3T3 and human diploid fibroblasts can also be used. Other lines, such as the A-31 clone of 3T3, are inefficient. Mitomycin C can be employed as an alternative to γ-irradiation to prevent feeder layer multiplication.

Solutions and media

Mitomycin C: dissolve a 2 mg vial in 4 ml of phosphate-buffered saline (PBS) and store at 4 °C

3T3 culture medium (3T3-CM):

Dulbecco's Modified Eagle Medium (DMEM) (ICN Biomedicals, 12-332–54)

10% donor calf bovine serum

Glutamine 4 mM
Penicillin–streptomycin 50 IU/ml

Culture of 3T3 fibroblasts and preparation of the feeder layer

In order to provide an efficient feeder for keratinocytes, an original frozen vial of 3T3 cells should be thawed, cultivated, passaged once (1:10) and used to generate several frozen stocks ($1–2\times10^6$ cells/vial). To prepare feeders, each frozen stock should then be thawed and cultivated (as follows) for no more than 10–12 passages (approximately 2–3 months).

Protocol

1 Seed 3T3 fibroblasts in 3T3-CM ($1–2\times10^4$ cells/cm^2) and feed every 3 days. Incubate cultures in a humidified atmosphere of 5% CO_2 at 37 °C.
2 Trypsinise subconfluent cultures by incubating them with 0.025% trypsin/0.01% EDTA until cells detach, neutralise trypsin by adding 3T3-CM, harvest cells and passage them 1:5 to 1:10.
3 To prepare feeder-layers, trypsinise subconfluent cultures as described above, resuspend them in 3T3-CM and irradiate with 6000 rads (cobalt-60 or caesium-137).
4 If using mitomycin C, incubate subconfluent 3T3 for 2 h at 37 °C with DMEM containing 10 μg/ml mitomycin C. Wash thoroughly with PBS and trypsinise as described above.
5 Seed γ-irradiated or mitomycin-C-treated 3T3 fibroblasts in keratinocyte culture medium (KCM, see below) without EGF at a cell density of $2.5\times10^4/cm^2$. Keratinocytes can be seeded simultaneously or after fibroblast adhesion is complete (at least 2 h after 3T3 seeding). In the latter case keratinocytes should be added within 24 h.

Isolation of keratinocytes from skin biopsies

Normal skin samples (full thickness) are usually obtained from newborn foreskins or from plastic surgery. In view of the presence of stem cells in hair follicles, hairy body sites are preferable. After chemical sterilisation of the donor site, skin samples are drawn, immediately placed in ice-cold DMEM containing 10% fetal calf serum and antibiotics, kept at 4 °C and processed as soon as possible (within 24 h), as storage decreases the viability of epidermal cells. Two main approaches are employed to isolate keratinocytes from skin biopsies: (1) direct enzymatic digestion of intact skin, or (2) initial enzymatic separation of epidermis from the dermis followed by enzymatic release of cells from epidermis.

(1) The first system consists of finely mincing skin biopsies and then stirring them with trypsin/EDTA (see below). At regular time intervals, the supernatant containing cells is recovered. This procedure allows the separation of a large number of cells from small biopsies (usually 4×10^6 cells/cm^2).

(2) To separate the epidermis from the underlying dermis, skin biopsies are incubated with 0.025% trypsin/0.02% EDTA for 60–90 min at 37 °C; epidermis is then peeled off by means of forceps and further disaggregated by mechanical agitation (pipetting). An alternative and widely employed procedure for separating epidermis from the dermis involves the use of dispase (see below), a neutral protease known to cleave the basement membrane zone of the skin by acting as a type IV collagenase. Intact and viable epidermal sheets are thus obtained and can be further disaggregated with trypsin. The use of thermolysin instead of dispase has also been described.

We find the first procedure (trypsin/EDTA on full-thickness biopsies) to be the most reliable since it avoids the potential loss of epidermal stem cells.

Materials, solutions and media

Materials

Autoclaved curved scissors
Iris scissors
Forceps
No. 4 scalpel handles
No. 22 stainless steel blades
Celstir flasks (Wheaton, USA)
Dulbecco's phosphate-buffered saline, Ca^{2+} and Mg^{2+} free (PBS^-)

Trypsin/EDTA

Trypsin solution (0.1%)
PBS (20×) 50 ml
Glucose 1 g
Trypsin 1-300 1 g
Phenol red (1%) 1 ml
Penicillin–streptomycin 10 ml (final concentration 50 U/ml–50 μg/ml)

Distilled water to 1 litre
Final pH adjusted to 7.45 just before filtration with a solution of 2 M NaOH, about 1 ml
Distribute in aliquots and store at −20 °C

EDTA solution (0.02%)
PBS (20×) 50 ml

Na_2EDTA 0.02 g
Distilled water to 1 litre.
Store at 4 °C.

A 1:1 mixture of the above solutions is used to isolate keratinocytes from skin biopsies and to trypsinise keratinocyte cultures.

Dispase solution
Dissolve 500 mg of dispase II in 100 ml Hank's Balanced Salt Solution (HBSS), containing 20 mM HEPES, pH 7.2, filter sterilise and store in aliquots at −20 °C.

Isolation of keratinocytes from full-thickness skin biopsies

Protocol

1 Place the skin sample in a petri dish with PBS^- and remove as much connective tissue as possible.
2 Wash with PBS^- and transfer the skin sample to a new petri dish. Add 2–5 ml trypsin and mince very finely with iris scissors and scalpel.
3 Transfer the minced biopsy into a Celstir flask, add trypsin/EDTA solution (3–4 ml/cm^2 of skin) and stir (gently) at 37 °C.
4 Harvest supernatant containing dissociated cells every 30 min and replace it with fresh trypsin/EDTA solution. Repeat this procedure until no more cells are recovered.
5 To neutralise trypsin activity, add an equal volume of serum-containing medium to the supernatant harvested at each trypsinisation. Centrifuge cells at 200 *g* for 5 min.
6 Resuspend the cell pellet in KCM without EGF (see below) and count cells in a haemocytometer.

Isolation of keratinocytes after dispase treatment

Protocol

1 Prepare skin samples by removing as much connective tissue as possible or, for large samples, keratomize (cut with a dermatome; Stryker Instruments, Kalamazoo, MI) to obtain split-thickness skin slices.
2 Incubate skin samples with 0.5% dispase for 60 min at 37 °C.
3 Separate epidermis from the dermis with forceps, then incubate epidermal sheets in 0.25% trypsin/0.02% EDTA solution for 15–20 min at 37 °C.

4 Dissociate epidermal cells by pipetting for 5 min and then continue as described above, points 4–6.

Preparation of keratinocyte culture medium

Keratinocyte culture medium (KCM)

DMEM and Ham's F-12 media (3:1 mixture)
10% fetal calf serum (FCS)*
Glutamine 4 mM
Penicillin–streptomycin 50 U/ml–50 μg/ml
Additives

* Each batch of serum should be tested for the ability to support good colony formation (CFE assay: see Introduction and Barrandon & Green, 1987): inoculate cultures with 100 and 1000 keratinocytes of a standard epidermal strain on lethally irradiated 3T3 cells (see below for the cultivation method). After fixation and rhodamine staining at 12 days, the number and size of colonies are compared with those generated using the standard serum. Note that only 10–20% of serum batches tested are suitable for keratinocyte cultivation.

Additives for keratinocyte medium

1 Insulin

Materials

Insulin 250 mg
HCl 0.005 M

Protocol

1 Dissolve 250 mg of insulin in 50 ml of 0.005 M HCl.
2 Distribute in 1 ml aliquots and freeze.
Add 1 ml to 1000 ml of complete medium (final concentration 5 μg/ml).

2 Hydrocortisone

Materials

Hydrocortisone 25 mg
Ethanol 95% 5 ml
DMEM 48 ml

Protocol

1 Dissolve 25 mg of hydrocortisone in 5 ml 95% ethanol; store at 4 °C.
2 Take 2 ml of stock, and make up to 50 ml with DMEM.
3 Distribute into 1 ml aliquots and freeze.
Add 2 ml to 1000 ml of complete medium (final concentration 0.4 μg/ml).

3 Cholera toxin

Materials

Cholera toxin 1 mg
Distilled water 1.18 ml
DMEM/10% FCS 10 ml

Protocol

1 Add 1.18 ml distilled water to 1 mg of cholera toxin (concentrated stock 10^{-5} M); store at 4 °C.
2 Add 0.1 ml of concentrated stock to 10 ml DMEM/10% FCS.
3 Distribute into 1 ml aliquots and freeze (10^{-7} M).
Add 1 ml to 1000 ml of complete medium (final concentration 10^{-10} M).

4 Epidermal growth factor (EGF)

Materials

EGF-human recombinant (Austral Biologicals) 200 mg
HCl 10 mM 20 ml

Protocol

1 Dissolve EGF (200 mg) in 20 ml HCl 10 mM.
2 Distribute into 1 ml aliquots and freeze.
Add 1 ml to 1000 ml of complete medium (final concentration 10 ng/ml).

5 Triiodothyronine

Materials

Triiodothyronine 13.6 mg
NaOH 0.02M 2 ml
Distilled water
Phosphate-buffered saline (PBS)

Protocol

1 Dissolve 13.6 mg in 2 ml NaOH 0.02M.
2 Make volume up to 50 ml with distilled water.

3 Take 0.25 ml and make up to 50 ml with PBS.
4 Distribute into 1 ml aliquots and freeze.
Add 1 ml to 1000 ml of complete medium (final concentration 2×10^{-9} M).

6 Adenine

Materials

Adenine 0.486 g
Distilled water 200 ml
HCl 37% 1 ml

Protocol

1 Dissolve 0.486 g of adenine in 200 ml of distilled water with stirring, add slowly 1 ml HCl 37% (v/v).
2 Distribute into 1 ml aliquots and freeze.
Add 10 ml to 1000 ml of complete medium (final concentration: 1.8×10^{-4} M).

Cultivation of keratinocytes

Primary cultures

Protocol

1 Suspend keratinocytes isolated from the skin biopsy in KCM without EGF.
2 Inoculate keratinocytes onto feeder layers at a cell density of 2.5×10^4/cm^2 for regular cultures. Inoculate 1000 cells/100 mm petri dish for CFE assays (see above). (Keratinocytes and lethally irradiated 3T3 cells can be seeded simultaneously. Alternatively, irradiated 3T3 cells can be seeded 2–24 h before keratinocyte seeding.)
3 Culture in 5% CO_2 and humidified atmosphere.
4 Use complete KCM (with EGF) from the first change of medium, 3 days after seeding. Change the medium every other day.

Keratinocyte sub-cultivation

Subconfluent primary cultures (7–9 days after plating) can be serially passaged or stored in liquid nitrogen.

Protocol

1 After rinsing with DMEM, incubate subconfluent primary cultures in trypsin/EDTA (10 ml/75 cm^2 flask) for approximately 20–30 min (until cells are detached and rounded up).

2 Transfer the cell suspension to a conical tube containing an equal volume of KCM, centrifuge at 200 *g* for 5 min and resuspend in KCM without EGF.
3 Inoculate keratinocytes onto feeder layers at a cell density of $5–6\times10^3/cm^2$ for regular cultures. Inoculate 100 cells/100 mm petri dish for CFE assays. Culture in 5% CO_2 and humidified atmosphere.
4 Use complete KCM (with EGF) from the first change of medium, 3 days after seeding. Change the medium every other day. Regular cultures will be subconfluent after 6–7 days.

The procedure can be repeated several times until senescence. Cell ageing can be monitored by evaluation of the clonogenic potential and the increase in the percentage of aborted colonies (see Introduction).

To maintain cell stocks, cells can be cryopreserved. Cells are resuspended in KCM (without EGF) containing 10% glycerol and transferred to suitable tubes or ampoules (2×10^6 cells/ml). Freezing can be performed with the aid of a programmable temperature controller using the following programme (alternative freezing procedures can also be used): −5 °C/min to +3 °C, −1 °C/min to −7 °C, −25 °C/min to −40 °C, +15 °C/min to −25 °C, −2 °C/min to −40 °C, −3 °C/min to −100 °C. Tubes are then transferred to liquid nitrogen.

Preparation of cohesive epidermal sheets

Dispase solution

Materials

Dispase II 250 mg
DMEM 100 ml

Protocol

1 Dissolve 250 mg of dispase II in 100 ml of DMEM with gentle stirring.
2 Rinse confluent keratinocyte cultures with DMEM.
3 Incubate in dispase solution until the epidermal sheet is detached.

Cultivation of keratinocytes in serum-free medium

The first and most widely used serum-free medium for growing human keratinocytes was described by Ham and coworkers in the early 1980s (Boyce & Ham 1985) and named MCDB 153. This medium sustains keratinocyte proliferation and suppresses the growth of human fibroblasts. Keratinocytes can thus be grown in the absence of a feeder layer of irradiated fibroblasts and, according to literature, clonal growth is achieved if MCDB 153 is supplemented with insulin, EGF, hydrocortisone, ethanolamine, phosphoethanolamine (defined MCDB 153). Bovine pituitary extract is usually added (complete MCDB 153) to initiate primary cultures, for frozen storage and for serial culture, and improves the number of cell doublings (up to 40). In order to obtain optimal cell proliferation, the Ca^{2+} concentration in MCDB 153 is lower (0.1–0.3 mM) than in standard media (1.2–1.8 mM): under these conditions keratinocytes form a monolayer of homogeneous, cuboidal cells but do not stratify. Increasing the Ca^{2+} concentration to 1.0 mM results in the induction of terminal differentiation, stratification and a reduction in keratinocyte growth rate. Several defined media analogous to MCDB 153 ('MCDB 153-derived') are commercially available and widely employed for keratinocyte cultures.

Media

KGM BulletKit, keratinocyte basal medium and supplements (Clonetics Corporation, San Diego, CA, USA)

or

KGM, complete keratinocyte growth medium (Clonetics)

Protocol

1 Isolate keratinocytes as described above, but no serum-containing medium should be employed to neutralise trypsin since its use reduces the subsequent passaging capability of keratinocytes.
2 Seed isolated keratinocytes ($1–3\times10^4$ cells/cm^2) in KGM. Change medium after 24 h and then three times weekly.
3 Incubate cultures in a humified atmosphere of 5% CO_2 and 95% air at 37 °C.
4 Trypsinise subconfluent cultures with 0.025% trypsin/0.01% EDTA, dilute the resulting cell suspension with KGM (1:20), centrifuge at 200 *g* for 5 min and passage them.

In some protocols, primary keratinocyte cultures are grown on lethally irradiated 3T3 fibroblasts and then subcultured in KGM.

Skin organotypic cultures

Skin organotypic cultures have been established in order to obtain full epidermal stratification and stratum corneum formation *in vitro*. They consist

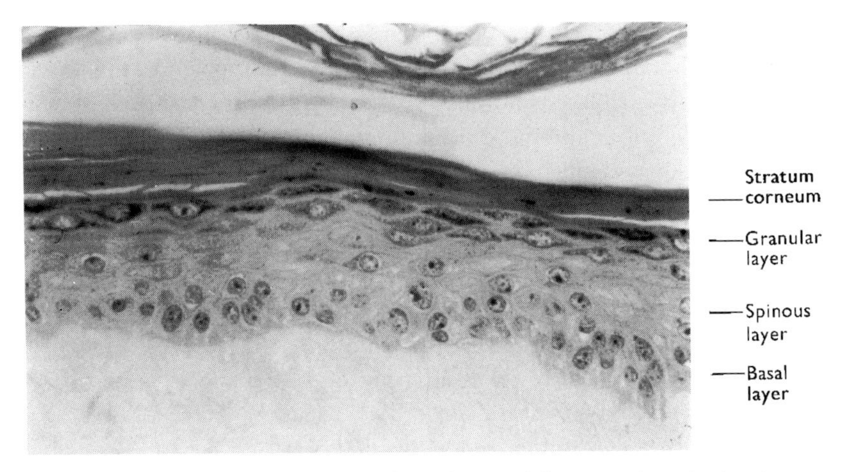

Fig. 7.4. Keratinocytes cultivated onto de-epidermised dermis as described in the text. Note the well-developed stratum corneum.

of seeding keratinocytes on a 'dermal' substrate and then lifting the whole culture at the air–liquid interface. The reconstituted epidermis displays a characteristic structure and morphology, i.e. the presence of distinct granular and cornified cell layers (Fig. 7.3), and expression and distribution pattern of terminal differentiation protein markers comparable to those of human epidermis *in vivo*. Two main types of dermal substrates have been described: (i) human de-epidermised dermis and (ii) mesenchymal cells incorporated into collagen gels/films.

The first approach, described by Pruniéras and collaborators in 1979, consists of human de-epidermised dermis (DED) as a substrate and isolated keratinocytes seeded on the top (Fig. 7.4). The dermal–epidermal unit is then lifted to the air–liquid interface. The human DED is subjected to repeated cycles of freezing and thawing and therefore does not contain living cells. A modification of this procedure envisages living DED placed beneath a collagen gel on which keratinocytes are plated.

A different skin organotypic culture model, devised by Bell and coworkers in 1979 and named 'dermal equivalent', has been widely used and modified over the years. In this system, living mesenchymal cells are incorporated into collagen gels (usually type I collagen) yielding a contraction of the latter. The extent of the contraction is proportional to the number of cells and inversely proportional to collagen concentration. Epithelial cells are plated on top of retracted gels and the whole unit is lifted to the air–liquid interface. Human dermal fibroblasts or established cell lines derived from them, murine 3T3 cells and also human dermal capillary endothelial cells can be incorporated in collagen gels. Air exposure is performed by placing the gels or gel holders on metal grids, glass rods, etc. In these models, an epidermis displaying all the cell layers of *in vivo* epidermis is obtained after 7–14 days of culture at the air–liquid interface. Culture devices and living skin equivalents are at present commercially available (e.g. from Organogenesis Inc. and Marrow-Tech).

De-epidermised dermis

Materials, solutions and media

Stainless steel rings (1 or 1.5 cm^2) and grids

Antimicrobial washing solution 1 (W1): Eagle's Minimum Essential Medium with Hank's salt solution (MEM–Hank's) without $NaHCO_3$, containing HEPES (20 mM), antibiotic/antimycotic solution (4×); (Gibco, supplied as 100×) (penicillin G sodium 100 U/ml, streptomycin 100 μg/ml, amphotericin B 0.25 μg/ml), gentamicin (3 mg/ml)

Antimicrobial washing solution 2 (W2): MEM-Hank's containing HEPES (20 mM), antibiotic/antimycotic solution (2×), gentamicin (1.5 mg/ml)

Antimicrobial washing solution 3 (W3): MEM-Hank's containing HEPES (20 mM), antibiotic/antimycotic solution (1×)

Skin organotypic culture medium (SOCM): Eagle's Minimum Essential Medium with Earle's salt solution (MEM–Earle), containing non-essential amino acids (1×), glutamine (2 mM), sodium pyruvate (1 mM), penicillin/streptomycin (100 IU/100 μg/ml), FCS 10%

Protocol

1 From large skin samples, prepare slices 1.5 mm thick using a dermatome and keep them in W2 solution at 4 °C until processed.
2 Cut the skin slices in 6×2 cm strips and wash them sequentially in W1, W2 and W3 solutions, 10 min each.
3 Float the skin slices, dermis-side down, on PBS^- in petri dishes and incubate them at 37 °C for 8–10 days; PBS^- should be changed every 3 days.
4 Remove the epidermis with fine forceps and cut the DED in squares of the appropriate size.
5 Place the DED in petri dishes so that the side that normally faces the epidermis is up, and freeze (−80 °C) and thaw (37 °C) them; repeat this cycle ten times. Store at −80 °C.

Skin organotypic cultures on DED

Keratinocytes isolated from skin biospies are cultivated on lethally irradiated 3T3 cells as described above. Subconfluent primary or secondary cultures are then trypsinised as above and seeded onto the de-epidermised dermis as follows:

Protocol

1 Thaw pieces of DED and transfer each piece with forceps into a 60 mm petri dish, maintaining the papillary dermis up.
2 Place a stainless steel ring on the DED and press the edge of the ring with forceps to ensure its adherence to the DED. Pour 5 ml medium outside the ring and incubate overnight at 37 °C.
3 Seed cultured keratinocytes (5×10^5 cells in 250 ml medium/cm^2 ring surface) inside the ring and incubate overnight to allow cell attachment.
4 Place a stainless steel grid in a new petri dish and add 6 ml medium.
5 Remove the rings and carefully transfer with forceps the whole cultures on the grid in order to lift the organotypic culture to the air–liquid interface. The level of the medium should not be higher than the level of the grid and air bubble formation must be avoided. Cultures are incubated in a humified atmosphere of 5% CO_2 at 37 °C and fed with fresh medium three times a week.

Dermal equivalent

The skin equivalent contains a dermal equivalent and a fully differentiated epidermis. The dermal equivalent consists of a network of type I collagen populated by dermal fibroblasts: collagen is mixed with culture medium, and the acidity of the collagen solution is neutralised with NaOH. Fibroblast suspension is then added, the mixture polymerises rapidly and fibroblasts are homogeneously distributed within the collagen matrix. Fibroblasts contract the collagen fibrils and give rise to the dermal equivalent.

The fully stratified epidermis is obtained by plating epidermal keratinocytes onto the surface of the dermal equivalent. Keratinocytes isolated from skin biospies are cultivated on lethally irradiated 3T3 cells as described above. Subconfluent primary or secondary cultures are then trypsinised as above and seeded onto the dermal equivalent.

Collagen solution

Materials

Rat tails from Wistar rats
Ethanol 70%
Acetic acid 0.1% (w/w)
NaOH 1M

Protocol

1 The minced tendons, cleaned with ethanol 70% (w/w), in acetic acid 0.1% (w/w) using 250 ml/g tendons. Stir this mixture for 48 h at 4 °C, then centrifuge, and neutralise the supernatant to pH 7 with 1 M NaOH.
2 After 24 h, precipitate the collagen and centrifuge; discard the supernatant and add 125 ml of 0.1% acetic acid/g of tendon to dissolve the collagen.
3 Store the solution at 4 °C; protein concentration is determined by the method of Lowry. The concentration of collagen is approximately 1 mg/ml in 0.05% acetic acid.

Preparation of collagen gels

Materials

Freshly isolated fibroblasts (or immortalised fibroblast cell lines), cultivated as described above
Fibroblast suspension in DMEM supplemented with 10% FCS
Sufficient 1M NaOH to neutralise
Collagen stock solution
DMEM 2×, supplemented with 20% FCS

Protocol

1 Form the collagen gel by mixing, at 4 °C, 0.25 ml of collagen stock solution, 0.125 ml NaOH and 0.375 ml DMEM 2×, containing 20% FCS.
2 Mix the solution well by swirling and pour into a bacteriological dish (about 8 cm^2). Formation of a gel starts within 1 min of the dishes being placed at 37 °C.

Construction of skin equivalent

Protocol

1 Place the gel on a nylon gauze on a stainless steel mesh support. Place stainless steel washer on the collagen gel (internal volume 0.1 ml).
2 Seed epidermal cells (3×10^5/0.1 ml) onto the surface of the collagen gel, and add keratinocyte medium to the dish so that the gel (but not keratinocytes) is covered by the culture medium.

After 2 weeks a fully stratified epithelium is formed at the air–liquid interface.

References

Barrandon, Y. & Green, H. (1987). Three clonal types of keratinocytes with different capacities for multiplication. *Proc. Natl. Acad. Sci. USA*, **84**, 2302–6.

Bell, E., Iversson, B. & Merrill, C. (1979). Production of tissue-like structure by contraction of collagen lattices by human fibroblasts of different proliferative potential *in vitro*. *Proc. Natl. Acad. Sci. USA*, **76**, 1272–8.

Boyce, S.T. & Ham, R.G. (1985). Cultivation, frozen storage, and clonal growth of normal human keratinocytes in serum-free media. *J. Tissue Cultivation Methods*, **9**, 83–93.

De Luca, M., D'Anna, F., Bondanza, S., Franzi, A.T. & Cancedda, R. (1988). Human epithelial cells induce human melanocyte growth *in vitro* but only skin keratinocytes regulate its proper differentiation in the absence of dermis. *J. Cell Biol.*, **107**, 1919–26.

De Luca, M., Tamura, R.N., Kajiji, S., Bondanza, S., Rossino, P., Cancedda, R., Marchisio, P.C., & Quaranta, V. (1990). Polarized integrin mediates human keratinocyte adhesion to basal lamina. *Proc. Natl. Acad. Sci. USA*, **87**, 6888–92.

Gallico, G.G., O'Connor, N.E., Compton, C.C., Kehinde, C. & Green, H. (1984). Permanent coverage of large burn wounds with autologous cultured human epithelium. *N. Eng. J. Med.*, **311**, 448–51.

Green, H. (1980). The keratinocyte as differentiated cell type. *Harvey Lect.*, **74**, 101–39.

Green, H., Kehinde, O. & Thomas, J. (1979). Growth of cultured human epidermal cells into multiple epithelia suitable for grafting. *Proc. Natl. Acad. Sci. USA*, **76**, 5665–8.

Hotchin, N.A., Gandarillas, A. & Watt, F.M. (1995). Regulation of cell surface β integrin levels during keratinocyte terminal differentiation. *J. Cell Biol.*, **128**, 1209–19.

Pruniéras, M., Régnier, M. & Schlotterer, M. (1979). Nouveau procédé de culture des cellules épidermiques humaines sur derme homologue ou hétérologue: préparation de greffons recombinés. *Ann. Chir. Plast.*, **24**, 357–62.

Rheinwald, J.G. & Green, H. (1975). Serial cultivation of strains of human epidermal keratinocytes: the formation of keratinizing colonies from single cells. *Cell*, **6**, 331–44.

Rochat, A., Kobayashi, K. & Barrandon, Y. (1994). Location of stem cells of human hair follicles by clonal analysis. *Cell*, **76**, 1063–73.

Tseng, H. & Green, H. (1994). Association of basonuclin with ability of keratinocytes to multiply and with absence of terminal differentiation. *J. Cell Biol.*, **126**, 495–506.

Watt, F.M. (1989). Terminal differentiation of epidermal keratinocytes. *Curr. Opin. Cell Biol.*, **1**, 1107–15.

Zambruno, G., Marchisio, P.C., Marconi, A., Vaschieri, C., Melchiori, A., Giannetti, A. & De Luca, M. (1995). Transforming Growth Factor-β1 modulates β1 and β5 integrin receptors and induces the *de novo* expression of the αvβ6 heterodimer in normal human keratinocytes: implication for wound healing. *J. Cell Biol.*, **129**, 853–65.

8

Skin gland and appendage epithelial cells

Terence Kealey, Michael Philpott, Robert Guy and Nick Dove

Introduction

The cell culture of the human skin glands and appendages has lagged behind that of interfollicular keratinocytes and fibroblasts because of problems of accessibility. Until recently, with the partial exception of the hair follicle, methods had not been described for the isolation of the skin glands and appendages. Over the last decade, however, techniques for the isolation, organ maintenance and cell culture of all the major human skin glands and appendages have been developed, and these have been valuable in the study of diseases including cystic fibrosis (eccrine sweat glands), hidradenitis suppurativa (apocrine sweat glands), the alopecias (hair follicles) and acne (sebaceous glands and the infundibulum of the sebaceous pilosebaceous follicle).

In the Introduction to this chapter we shall briefly review, under the heading of Anatomy, the major human skin glands and appendages. We shall then review, under the appropriate headings, their embryology, cell physiology and histology and cell kinetics where these affect their cell culture. In the Methods section we shall successively review each gland or appendage. Their isolation will first be described, followed by a brief note on their organ maintenance, and then a description of the culture of their different cells.

Anatomy

There are four major skin glands or appendages, namely the hair follicle, the sebaceous gland, and the eccrine and apocrine sweat glands.

The pilosebaceous unit

The hair follicles and sebaceous glands are generally associated as pilosebaceous units, each unit consisting of one follicle and one gland, though isolated sebaceous glands have been reported in some non-skin epithelia such as the oesophagus. The skin of almost the whole of the human body contains pilosebaceous units, only the palms of the hands, the soles of the feet and the penis being free of them. Pilosebaceous units are characteristic of mammals, and different species demonstrate different classes of unit. Whiskers, for example, are not seen in the human. In the human, there are at least four classes of pilosebaceous unit:

The terminal pilosebaceous unit is found on the scalp of both sexes, and in the beard area of normal adult males. It constitutes a large terminal hair and a large sebaceous gland.

The apo-pilosebaceous unit is found in the axillary and pubic areas of normal adults of both sexes. It is similar to the terminal pilosebaceous unit, except that one or more apocrine sweat ducts feed into the perishaft space above the opening of the sebaceous gland. The transformation of this unit from a vellus one is mediated by sex hormones at puberty.

The vellus pilosebaceous unit is found in areas of the skin that appear to the naked eye to be essentially hairless. It comprises a small hair and a small sebaceous gland. The vellus unit of the beard area in particular, but also of the chest, back and limbs, can be converted into terminal units under the influence of androgens and other sex hormones.

The sebaceous pilosebaceous unit is found on the face, chest and back. It is the seat of acne, and only develops at puberty under the influence of sex hormones. It comprises a large sebaceous gland, a small hair follicle and a large infundibulum (opening to the surface of the skin).

Sweat glands

Sweat glands, which are characteristic of mammals, secrete a mixture of fluid and lipid. Many mammals possess only one type of sweat gland, but those of the human are the most specialised in that two types are found: the eccrine sweat gland which secretes fluid, and the apocrine gland which largely secretes lipid. (A third type, the apoeccrine, a mixture of the two, has also been described. It is found in the axilla and will not be discussed further here.)

Eccrine sweat glands are found all over the skin. The average human possesses 1–2 million. The glands secrete hypotonic fluid on to the surface of the skin. The function of most eccrine sweat glands is thermoregulation and the evolution in *Homo sapiens* of such specialised eccrine glands may be an exploitation of his relatively hairless state. The eccrine sweat glands of the palms of the hands and the soles of the feet secrete fluid to strengthen grip. Their secretion, therefore, is stress-mediated.

The unusual evolution of the human eccrine sweat glands is reflected in their innervation. Like other mammalian sweat glands they are controlled by the sympathetic nervous system, but their major secretagogue is cholinergic (muscarinic) rather than (β) adrenergic as in other mammals.

Apocrine sweat glands develop at puberty under the influence of androgens and other sex hormones. They are restricted to the axilla and inguinal region. Only 10% of them secrete directly on to the skin surface; 90% feed into the perishaft space of apo-pilosebaceous units.

Other skin glands and appendages

Other skin glands and appendages such as the ceruminous gland (a modified apocrine sweat gland that secretes wax into the external auditory meatus) or the meibomian gland (a modified sebaceous gland of the eyelids) or the mollian gland (a modified sweat gland of the eyelids) have not been cell cultured and will not be discussed here.

Embryology

The skin glands and appendages share a similar embryology, which will be described here as it determines much of their cell culture.

Around the third month of human fetal life, epidermal cells cluster together over aggregations of dermal fibroblasts. From these foci, epidermal cells grow down into the dermis to give rise to the different skin glands and appendages. It remains unclear whether the epithelial cells direct the mesenchymal cells to aggregate or vice versa. This has been most intensively studied in the hair follicle (which has a prominent mesenchymal element) where the epidermal cells appear to cluster prior to the aggregation of the mesenchymal cells, so it is assumed that they initiate the process of pilosebaceous formation. Recombination experiments suggest, however, that thereafter the mesenchymal cells direct the subsequent growth and differentiation of the pilosebaceous unit. Cultured dermal papilla cells, when introduced within adult rat footpads, can induce the formation of hair follicles, which confirms that the dermal papilla may contain all the information required for hair follicle embryogenesis. Such adult dermal papilla-induced hair follicles, however, lack a sebaceous gland, which indicates that the sebaceous gland's development is autonomous of the dermal papilla.

The chemical factors that regulate the embryological development of the skin glands and appendages remain unknown, but epidermal growth factor

(EGF) has emerged as important from the study of *Ta* mutant mice. The *Ta* mutation is the murine equivalent of anhidrotic ectodermal dysplasia, an inherited X-linked human disease characterised by an absence, or a scarcity, of pilosebaceous units, sweat glands, teeth, and other skin glands and appendages. Further, it has been shown that the injection of EGF into *Ta* mice will correct the disease and induce sweat gland development. Moreover, mice whose EGF receptors have been inactivated transgenically will demonstrate, amongst other pathologies, sparse hair.

The homeobox genes *Msx*-1 and *Msx*-2 have recently been identified within growing feather bud epithelia, but the homeobox genes associated with mammalian skin gland and appendage development have yet to be described.

Overview of cell physiology

Pilosebaceous unit

The pilosebaceous unit possesses two populations of rapidly dividing epithelial cells:

Hair matrix cells. The matrix cells at the base of the hair follicle divide to generate daughter cells that, after a transit time of around 2.5 days, terminally differentiate and die, having synthesised the keratins and other proteins that largely constitute the hair fibre.

Sebocytes. The sebocytes of the sebaceous glands arise from basal cells, and they also terminally differentiate and die with a transit time of around 3 weeks, to release their cytoplasmic lipid as sebum. This suicidal form of secretion is described as holocrine.

Stem cells

Label-retaining, histological, clonogenic, keratin 19 expression and other experiments indicate that the epithelial stem cells of the pilosebaceous unit are located within the upper outer root sheath (ORS). Further, it appears that these also act as stem cells for the epidermis, as ORS keratinocytes can demonstrate phenotypic plasticity both *in vivo* and *in vitro*. Should the epidermis be stripped *in vivo* it can be re-epithelialised by ORS cells; should ORS keratinocytes be seeded on to a living skin equivalent *in vitro*, they will demonstrate an epidermal phenotype (Limat *et al.*, 1991). It has been suggested that vellus hairs have been retained by the relatively hairless *Homo sapiens* to locate the epidermal stem cells safely, deep in the skin.

Sweat glands

In contrast to the pilosebaceous unit, secretion by the sweat glands is not holocrine. Indeed, to the best of our knowledge, no mitotic figures have been reported for sweat glands *in situ*, with the exception of the cells of the eccrine (and 10% of apocrine) duct infundibulae, which divide to migrate *pari passu* with epidermal keratinocytes.

Early histological studies suggested that apocrine glands secrete in an apocrine fashion (hence the name); i.e. the apices of the secretory cells, which are rich in lipid-containing granules, are shed into the lumen. Early histology also suggested that eccrine sweat glands secrete in an eccrine fashion; i.e. fluid passes between the cells which suffer no rupture. More recent histological studies have suggested that both exocytosis and eccrine secretion may also occur in the apocrine sweat gland, and that exocytosis may be seen in the eccrine sweat gland dark cells. The precise nature of sweat gland secretion, therefore, remains uncertain.

Interestingly, the scarcity of sweat gland mitoses *in situ* does not prevent their successful cell culture *in vitro*, which indicates that sweat gland cells are in G_0 rather than terminally differentiated like those of the hair follicle inner root sheath (IRS) or the epidermal stratum corneum.

Overview of histology and cell kinetics

The hair follicle

The hair follicle is composed of dermal components (the dermal papilla and connective tissue sheath) and epidermal components (the matrix, IRS and ORS: see Fig. 8.1).

The dermal papilla and connective tissue sheath. The dermal papilla is located at the base of the hair follicle and consists of specialised dermal fibroblasts surrounded by an extensive extracellular matrix that is rich in basement membrane proteins, fibronectin and proteoglycans.

The matrix and ORS. Surmounting and surrounding the dermal papilla are the matrix cells, which [^{3}H]methylthymidine autoradiography has shown have a very rapid rate of cell division. [^{3}H]methylthymidine pulse chase has shown that, as the matrix cells divide, they undergo an upward migration giving rise, respectively, to the hair cortex (which keratinises to form the hair fibre), the cuticle (the outermost layer of the hair fibre), the IRS (which is believed to provide structural support for the developing hair fibre) and the innermost cells of the ORS.

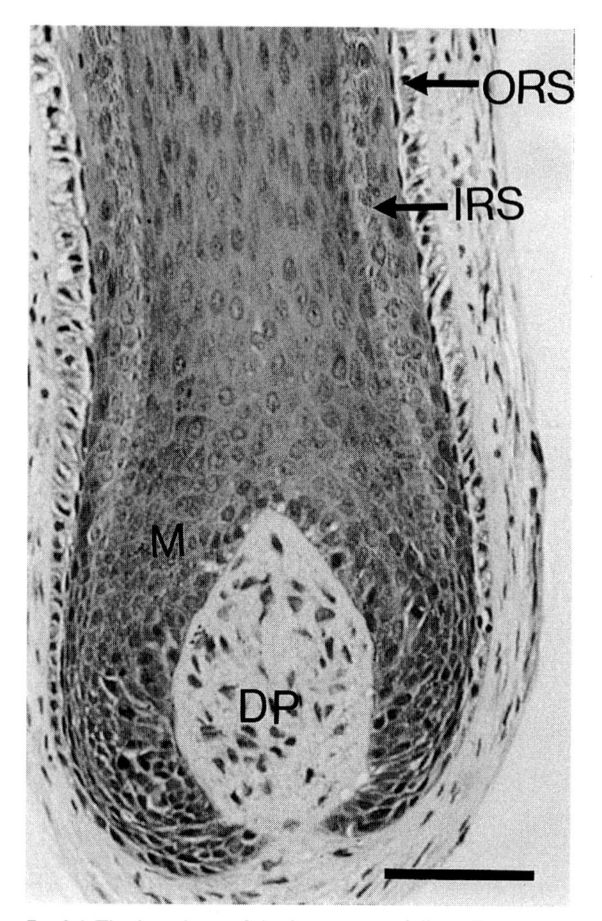

Fig. 8.1. The histology of the human hair follicle. This shows the bulb (growing) area of a hair follicle that has been isolated by modified microdissection and fixed and stained conventionally for light microscopy. The dermal papilla (DP), matrix cells (M), inner root sheath (IRS) and outer root sheath (ORS) are all visible. The scale bar represents 100 μm.

The ORS possesses its own basal cells which sit on the basement membrane and which divide *in vivo*. It is believed that these give rise to the outermost cells of the ORS. It is now also thought that many of the basal dividing ORS cells represent daughter transient amplifying cells of the putative stem cells of the upper ORS, as do the matrix cells, but the precise kinetics of the movement of cells between these three components has yet to be fully determined (Cotsarelis *et al.*, 1990).

The eccrine sweat gland

The human eccrine sweat gland consists of two parts: a secretory coil and a collecting duct. The primary secretion, from the secretory coil, is effectively a plasma ultrafiltrate. The collecting duct actively reabsorbs about half of the NaCl from the primary secretion during its passage.

The secretory coil consists of three cell types: the myoepithelial cells, the clear cells and the dark cells. All three cell types abut on to the basement membrane, but only the clear and dark cells abut luminally.

It is assumed that the myoepithelial cells provide structural support and expulsive propulsion to fluid secretion, but they cannot be essential to fluid secretion since at least one exocrine gland, the pancreas, lacks them. The dark cells are assumed to secrete mucus, and they are not considered essential to eccrine fluid secretion since at least one eccrine sweat gland, that of the rat paw, lacks them. It is thus assumed that fluid is secreted by the light cells. During active fluid secretion the intercellular spaces and canaliculi that bathe the clear cells dilate, which is assumed to reflect the basolateral ion and water transport common to all exocrine glands.

Dye injection experiments have shown that the individual cells of the secretory coil are linked to the other cells of the same type in a syncytium, but that the three cell-specific syncytia are not linked.

The collecting duct of the human eccrine sweat gland consists of two cell layers, linked as one syncytium. The inner cell layer is packed with tonofilaments, and the density of these increases along the duct; these are assumed to provide structural support against the osmotic swelling that would result from the hypotonic sweat. The outer cell layer is packed with mitochondria, whose density decreases down the duct, and these are assumed to provide the energy for NaCl reabsorption which is maximal proximally.

The apocrine sweat gland

Like the eccrine gland, the human apocrine sweat gland comprises a secretory coil and a collecting duct. The secretory coil, however, comprises only two cell types: myoepithelial and secretory. The histology of the collecting duct has not been comprehensively described.

The sebaceous gland

There is only one cell type in the sebaceous gland, the sebocyte, which progresses from the basal undifferentiated sebocyte (a transient amplifying cell) through the partially differentiated sebocyte (which starts to accumulate lipid) to the fully differentiated sebocyte (which is congested with lipid droplets). This progression is generally associated with a migration into the centre of the gland, where the mature differentiated sebocyte bursts and dies, to liberate its sebum via the sebaceous duct into the perishaft space or infundibulum, and then onto the skin surface.

The sebaceous infundibulum

The sebaceous infundibulum is continuous with the epidermis, of which it is a downgrowth. In the deeper layers of the dermis, the strata corneum and granulosa of the infundibulum become thinner, but otherwise the histology of the infundibulum resembles that of the epidermis. Like the epidermis, its basal transient amplifying cells are located on the basal lamina. Their transit time has not been described.

Materials

Abbreviations
BSA: bovine serum albumin
EBSS: Earle's Balanced Salt Solution
DMEM: Dulbecco's Modified Eagle Medium
EDTA: ethylenediaminetetraacetic acid
EGF: epidermal growth factor
PBS^+: phosphate-buffered saline with 2 mM Ca^{2+} and 2 mM Mg^{2+}
PBS^-: phosphate-buffered saline without Ca^{2+} or Mg^{2+}

Media and supplements, which must be of tissue culture grade, may generally be purchased from any reputable company that supplies tissue culture products. For some products, however, we will indicate which company we have found optimal.

Media

Williams E (including phenol red-free), DMEM, EBSS and PBS we buy as liquid media (1×). In our experience, medium made up from powder sustains suboptimal growth, particularly of organs

Keratinocyte serum-free medium (keratinocyte SFM; GIBCO 17005-034) is supplied as a liquid basal medium (1×). GIBCO supply this with two supplements: EGF (37000-015) and bovine pituitary extract (37000-015). For the use of these supplements see below

Mammary Epithelium Basal Medium is supplied by Clonetics (CC-3151)

Supplements

Penicillin and streptomycin. Bought as a single solution containing both antibiotics at 100× concentration. Kept frozen, aliquots are thawed and 1 ml added to 100 ml of medium prior to use to achieve final concentrations of 100 U/ml penicillin, 100 μg/ml streptomycin

Amphotericin B. Supplied as a solution (100×). Store frozen and add at 1 ml per 100 ml medium to achieve 2.5 μg/ml

L-Glutamine. Bought as liquid medium at 200 mM (100×). Stored frozen in aliquots of 10 ml, 1 ml is added to 100 ml tissue culture medium, where necessary, immediately prior to use, to achieve a final concentration of 2 mM. (All tissue culture media should routinely contain 2 mM glutamine, but some manufacturers now supply some media already supplemented with glutamine)

$CaCl_2(10H_2O)$. Used to supplement keratinocyte SFM to a final Ca^{2+} concentration of 2 mM. Dissolve 2.79 g $CaCl_2(10H_2O)$ in 100 ml keratinocyte SFM, filter sterilise if not of tissue culture grade, and add 1 ml per 100 ml of medium. Store at 4 °C

Hydrocortisone. Bought as vials of 1 mg powder, to which 1 ml of 100% ethanol is added, followed by 19 ml of the appropriate tissue culture medium. Store frozen in 1 ml aliquots and add 20 μl to 100 ml of Williams E immediately prior to use to give a final concentration of 10 ng/ml

Insulin. Bought as vials of 100 mg powder, to which 10 ml of sterile pure water is added, followed by 100 μl sterile glacial acetic acid. Store frozen in 100 μl aliquots and add 100 μl to 100 ml of tissue culture medium immediately prior to use to give a final concentration of 10 μg/ml

Transferrin. Supplied as 10 mg powder. Reconstitute with 10 ml of the appropriate tissue culture medium and store frozen in 1 ml aliquots. Add 1 ml per 100 ml medium, to give a final concentration of 10 μg/ml

Sodium selenite. Supplied as 1 mg powder. Reconstitute with 10 ml tissue culture medium and store frozen in 10 μl aliquots. Add 10 μl per 100 ml medium to give a final concentration of 10 ng/ml

Triiodothyronine. Supplied as 1 mg powder. Reconstitute with 1 ml 1 M NaOH, swirl until dissolved and add 49 ml tissue culture medium. Store frozen in 10 μl aliquots and add 10 μl per 100 ml medium to give a final concentration of 3 nM

1% (v/v) trace elements mix. Supplied as crystals. Add 10 ml distilled deionised water and store frozen. Add 1 ml per 100 ml medium

Bovine pituitary extract. Supplied as 5 mg powder. Reconstitute with 1 ml tissue culture medium. Store frozen and add 200 μl to 100 ml medium to give a final concentration of 10 μg/ml. For 50 μg/ml bovine pituitary extract, as recommended by Gibco for use with their keratinocyte SFM, use the dilutions supplied by the manufacturer

Epidermal growth factor (EGF). Supplied as a powder. Aliquots can be dissolved in PBS or tissue culture medium and stored frozen. For 5 ng/ml EGF as recommended by GIBCO for use with their keratinocyte SFM, use the dilutions supplied by the manufacturer. The standard tissue culture range of EGF concentrations is 2–20 ng/ml

Dispase (Boehringer Mannheim). Use at a working concentration of 2 U/ml. Dissolve the dispase in warmed complete medium

Methods

I The hair follicle

Isolation of human hair follicles

Plucking

Plucking is the oldest technique and it provides a good yield of viable ORS for primary cell culture. But plucking fails to recover the dermal papilla or the sebaceous gland, which are retained in the skin, and it so damages the follicle that plucked hair follicles will not grow hair on organ culture.

Microdissection

The obvious method of microdissection has been employed for many years. Biopsies of human skin are cut into thin strips, inverted so that the hypodermal side is facing up, and the hair follicles can then be easily visualised and microdissected. Although such preparations are suitable for cell culture, previous operators have so damaged the follicles that they failed to grow hair on subsequent organ maintenance. Recently, it was shown that meticulously gentle dissection, taking scrupulous care to avoid damaging the bulb, could isolate follicles capable of hair growth *in vitro*, though at suboptimal rates.

Modified (inverted) microdissection

Modified microdissection is a much better and much easier technique for the isolation of whole, undamaged hair follicles (Philpott *et al.*, 1990). It depends on cutting the skin at the interface between the dermis and the subcutaneous fat, and then pulling the follicles up out of the fat with forceps that grip the cut, upper part of the follicle, not the bulb, as it is trauma to the bulb that blocks hair growth *in vitro*. Follicles can be isolated from full-thickness skin at any site where the terminal hairs penetrate into the subcutaneous fat; these sites include scalp skin removed at facelift surgery, and the beard, axillary and pubic skin.

The skin is cut into thin strips (approximately 10 mm×5 mm) which are kept in EBSS:PBS$^+$ (with Ca^{2+} and Mg^{2+}) (1:1) until required for isolating hair follicles. This medium can be used on the open bench as it does not become too alkaline. Hair follicles are isolated by taking a strip of skin and, under a dissecting microscope, cutting very carefully through the skin at the dermo-subcutaneous fat interface using a sharp scalpel blade (Swann-Morton no. 24). This cuts the hair follicles just beneath the opening of the sebaceous gland, resulting in the lower portion of the hair follicle, including the bulb, being left in the subcutaneous fat.

The hair follicles are then easily isolated from the subcutaneous fat by placing the cut side upwards under the dissecting microscope. Using watchmaker's forceps (no. 5), carefully grasping the ORS of the cut end of the hair follicle, the hair follicle can be gently removed from the fat. Isolated hair follicles are placed in EBSS:PBS$^+$ (1:1) prior to being used for experiments. This method will yield several hundred isolated hair follicles within 2 h from a piece of skin 2 cm×1 cm. The yield of follicles from an individual piece of skin is variable depending upon the thickness of the skin and underlying fat, the size of the follicles and the depth at which they penetrate into the fat.

Skin should be used within a few hours of removal from the patient, though follicles isolated from skin left overnight in a refrigerator at 4 °C will still grow *in vitro*. Hair follicles should be left in isolation medium for the shortest time possible (<2 h). When hair follicles are being handled, only the upper ORS (the cut end) is held in the tweezers, and not the rounded hair follicle bulb.

Organ culture of isolated human hair follicles in serum-free defined medium

Isolated human hair follicles are maintained in individual wells of 24 well multiwell plates (Falcon 3047) containing 500 μl of Williams E medium supplemented with 2 mM L-glutamine, 10 μg/ml insulin, 10 ng/ml hydrocortisone, 100 U/ml penicillin and 100 μg/ml streptomycin at 37 °C in an atmosphere of 5% CO_2/95% air (Westgate *et al.*, 1993). Length measurements are made on isolated hair follicles using a Nikon Diaphot inverted microscope with an eyepiece measuring graticule (2.5 mm). Medium is usually changed every 3–4 days, although hair follicles will grow normally for up to 10 days without a medium change.

Primary culture of ORS keratinocytes

There are two basic methods that can be used for establishing ORS keratinocyte cultures: primary explants can be established from plucked or dissected follicles or keratinocytes can be isolated from the ORS by trypsin digestion of the hair follicle (Stark *et al.*, 1987) and then plated as single-cell suspensions onto 3T3 feeder layers. The primary explant from whole hair follicles is generally recommended for most purposes as it is easy, does not require 3T3 feeder layers (though these can be employed) and readily generates confluent cultures. But one theoretical advantage of the single-cell suspension method is that it will generate a confluent, homogeneous monolayer without passage. This will minimise de-differentiation.

Primary explant culture of ORS keratinocytes

Hair follicles are isolated either by dissection or plucking and placed under a dissecting microscope where, with a scalpel blade, the hair follicle bulb is removed and the remainder of the hair follicle cut into a number of small pieces (approximately 1 mm in length). Five to ten pieces of follicle are placed into 25 cm^2 flasks, directly onto tissue culture plastic, and covered with a thin layer of culture medium consisting of keratinocyte SFM (GIBCO) medium supplemented with 5 ng/ml EGF, 50 μg/ml bovine pituitary extract, and incubated at 37 °C in an atmosphere of 5% CO_2/95% air. After 24 h, most of the hair follicle pieces will have attached to the tissue culture plastic, and 5 ml of fresh medium can be added. Under these conditions, ORS keratinocytes readily explant.

ORS keratinocytes will explant in more conventional media, such as Eagle's Minimum Essential Medium (MEM), but these require supplementation with up to 15% fetal calf serum. They are, therefore, undefined, and they may promote fibroblast or other non-ORS cell overgrowth. The keratinocyte SFM medium is defined (though it is supplemented with 50 μg/ml bovine pituitary extract) and it does not require the addition of FCS. It promotes cell proliferation, and inhibits terminal differentiation, of keratinocytes, because it only contains 0.09 mM calcium, and it contains 5 ng/ml EGF. Primary cultures of keratinocytes will tend to differentiate in physiological concentrations of calcium (Hennings *et al.*, 1980) in the absence of EGF.

To promote attachment, some authors expose hair follicles to a short period of enzyme digestion such as 20 min in 1 mg/ml collagenase in PBS at 37 °C, before placing them on the plastic; others report that prior coating of the plastic with collagen improves keratinocyte explant, but neither of these interventions is obligatory. None of the older substrates, such as bovine eye lens capsules, that were once used for ORS keratinocyte explant, are now necessary.

ORS culture from single cell suspensions

Hair follicles, minus their bulbs, are incubated in 0.1% trypsin, 0.02% EDTA in PBS for 10 min at 37 °C. A single-cell suspension is obtained by vigorously pipetting the follicles in DMEM supplemented with 10% FCS. Dissociated keratinocytes are harvested by centrifugation (200 *g* for 10 min). The cells from one hair follicle are suspended in 2 ml of: 3 parts DMEM containing sodium pyruvate/1 part Ham's F-12, supplemented with 10% FCS, 5 μg/ml insulin, 0.4 μg/ml hydrocortisone, 0.135 mM adenine, 2 nM triiodothyronine, 10^{-10} M cholera toxin, 10 ng/ml EGF, 50 U/ml penicillin, 50 mg/ml streptomycin.

Keratinocytes are seeded at $2–3 \times 10^3$ cells/cm^2 on a 3T3 feeder layer. Cultures are incubated at 37 °C in 5% CO_2/95% air. Feeder layers are renewed weekly after selective removal of 3T3 fibroblasts with 0.02% EDTA.

Once primary cultures of ORS keratinocytes have been established, either from primary explants or from single-cell suspensions, they can be serially cultivated for two to four passages on 3T3 feeder layers.

3T3 feeder layers

3T3 feeder layers are generated from stocks of NIH 3T3 fibroblasts which can be obtained from the European Collection of Animal Cell Cultures (ECACC, Division of Biologics, PHLS Centre for Applied Microbiology and Research, Porton Down, Salisbury, SP4 OJG, UK; NIH-3T3 4-2; cat. no. 86041101). They are grown in DMEM supplemented with 10% FCS, 2 mM L-glutamine, 100 U/ml penicillin and 100 μg/ml streptomycin, and 2.5 μg/ml amphotericin B in 75 cm^2 flasks. Medium is changed every 3 days. On reaching confluence, cells are passaged using 0.05% trypsin, 0.53 mM Na-EDTA and plated out at 1.5×10^4 cells/cm^2.

Prior to use as a feeder layer, 3T3 cells must be lethally irradiated or exposed to mitomycin C. Irradiation is faster and easier, but depends on access to the appropriate equipment. The cells are trypsinised, resuspended in medium and irradiated in a sterile plastic container with 6000 rads from a ^{60}Co source. Then 75 cm^2 flasks are innoculated with 1.5×10^6 irradiated cells and incubated until the cells are attached (4–12 h). Keratinocytes are plated out at a density of 2.5×10^4 cells/cm^2.

Markers of ORS keratinocyte differentiation and phenotypic plasticity

ORS keratinocytes *in situ* express five cytokeratins: 5, 6, 14, 16 and 17. ORS keratinocytes grown *in vitro* on 3T3 feeder layers produce the same five cytokeratins, indicating a good retention of phenotype (Stark *et al.*, 1987). They will also develop multilayers *in vitro*, as *in situ*. But neither *in situ*, nor on 3T3 cells *in vitro*, do ORS keratinocytes express certain markers of interfollicular epidermal differentiation, including keratohyalin granules, 67 kDa keratin, involucrin, membrane-bound glutaminase, filaggrin or the development of a stratum corneum. But Limat *et al.* (1991) have shown that if ORS keratinocytes are seeded *in vitro* onto dermal equivalents, and are raised to the air/liquid interface, a histologically normal, fully differentiated epidermis develops, expressing those interfollicular markers of differentiation in staining patterns largely similar to those seen in the epidermis *in situ*. This *in vitro* phenotypic plasticity, therefore, largely mimics that seen *in vivo* when ORS keratinocytes re-seed a denuded epidermis. The only important difference in epidermal differentiation markers is that ORS-derived keratinocytes continue to express cytokeratins 5, 6, 14, 16 and 17 on dermal equivalents, but of those only 5 and 14 are normally expressed in epidermis *in situ*, though 6 and 16 are 'hyperproliferative' cytokeratins that are expressed *in situ* in hyperproliferative conditions (Stark *et al.*, 1987).

When Limat *et al.* (1991) published their findings that ORS keratinocytes seeded *in vitro* onto dermal equivalents could generate epidermal equivalents, they suggested that there were no significant differences in differentiation between those generated from interfollicular keratinocytes (see Chapter 7) and those generated from ORS-derived keratinocytes. Recently, however, they reported that ORS-derived keratinocytes express low levels of class 1 MHC antigens in culture, unlike those derived from interfollicular epidermis (Limat *et al.*, 1994). It is not clear whether this low level of expression has any significance *in vitro*.

Longer-term culture and passage

Primary human ORS cultures will survive up to nine passages (performed as described above for 3T3 feeder layers with trypsin/EDTA) with no loss of differentiation as determined by cytokeratin and filaggrin expression. After nine passages, however, ORS cell death was invariant. Crisis or spontaneous immortalisation was never observed.

Cryopreservation

In our experience, ORS keratinocytes can be successfully cryopreserved under liquid nitrogen using standard techniques (cells are suspended in FCS containing 10% dimethylsulphoxide, frozen at 1 °C/min and then stored in liquid nitrogen; followed by their rapid thawing and the slow readmission of medium).

Immortal ORS keratinocyte cell lines

To the best of our knowledge, no immortalisation of ORS keratinocyte cell lines has been attempted or achieved, though Kanazaki *et al.* (1983) reported a cell line derived from a human trichilemmoma, a benign ORS-derived tumour. These cells will grow *in vitro* on 3T3 feeder layers, and will form follicle-like aggregates on culture.

IRS culture

The IRS, on microdissection, does not explant or proliferate under a range of standard conditions in culture (Stark *et al.*, 1987). This is not surprising as it represents terminally differentiated cells that resemble those of the stratum corneum. Since matrix cells are so difficult to grow (see below), this confirms that the only source of epithelial cell outgrowth from the whole hair shaft is the ORS.

Stem cell culture

Hair follicle stem cells have never been isolated or cultured as such. However, it has been shown by clonal analysis that the ORS contains cells with high clonogenicity, indicating that ORS cultures may contain stem cells. But these have yet to be characterised *in vitro*.

The only *in vitro* culture model that has been characterised, in part, for stem cells is the organ-maintained whole hair follicle, where stem cells can be recognised by their specific staining of keratin-19, and where it has been shown that EGF will stimulate their clonal proliferation *in vitro*.

Germinative epithelium cell (of matrix cell) cultures

The human hair follicle germinative epithelial cells, which give rise to the matrix cells, can be cultured by the method of Reynolds *et al.* (1993) for the culture of germinative epithelium from rat vibrissae follicles. Matrix cell culture requires a dermal papilla fibroblast feeder layer, as more conventional feeder layers such as 3T3 fibroblasts fail to sustain the characteristically small and circular matrix cell phenotype on culture.

Dermal papilla fibroblast feeder layers

Dermal papilla fibroblast feeder layers are prepared as described by Messenger (1984). Hair follicles are isolated from human skin and placed in isolation medium as described above, but supplemented with 1% BSA, penicillin (100 U/ml) and streptomycin (100 μg/ml). The addition of BSA to the isolation medium reduces adhesion of the isolated dermal papilla and facilitates easy transfer for subsequent culture. Using a scalpel blade the lower hair follicle bulb, containing the dermal papilla, is removed by transecting the hair follicle through the keratogenous zone. The lower follicle bulb is then placed in fresh isolation medium.

Dermal papillae are isolated from the hair follicle bulbs as follows. First, using either fine dissecting needles (Watkins and Doncaster, Maidstone, Kent, UK), or 25 G Microlance syringe needles attached to 1 ml plastic syringes, the hair follicle matrix cells are removed. This is achieved by holding the follicle bulb to the bottom of the petri dish with watchmaker's forceps, then using a needle to remove the matrix cells from the cut end of the follicle bulb. These cells are easily removed; frequently, applying pressure to the base of the follicle bulb will result in the matrix being ejected from the cut end of the follicle. The remaining hair follicle bulb contains the connective tissue sheath and dermal papilla which sits centrally within the follicle bulb. The connective tissue sheath is then cut open with syringe needles and the dermal papilla isolated by careful dissection. This procedure is difficult but can easily be mastered with practice and up to 20 dermal papillae can be isolated within 2 h.

Once isolated, up to five dermal papillae are placed in individual wells of 24 well multiwell plates and covered with 100 μl of culture medium which consists of either MEM or Williams E medium supplemented with 2 mM L-glutamine, 20% FCS, penicillin (100 U/ml) and streptomycin (100 μg/ml). Cultures are maintained at 37 °C in an atmosphere of 5% CO_2/95% air. Within 48 h most dermal papillae will have attached to the tissue culture plastic, the old medium can be removed by gentle aspiration and 500 μl of fresh medium added. Medium is usually changed every 3 days. Initial growth is usually slow but within 7 days explants from the papilla can usually be seen under phase contrast microscopy. Primary explants grow slowly, and the cells usually aggregate, so they rarely achieve confluence; it is thus not necessary to arrest their growth prior to overlaying them with germinative epithelial cells. If required, dermal papilla fibroblasts may be passaged after 4 weeks.

Preparation and culture of matrix cells

The hair follicle matrix and germinative epithelium are removed from the hair follicle bulb as described above. Isolated germinative epithelial tissue is then placed in either 24 well multiwell plates or 35 mm petri dishes on feeder layers of human dermal papilla fibroblasts, and covered with culture medium consisting of MEM containing 20% FCS, 2 mM L-glutamine, 10 μg/ml EGF, 50 μg/m bovine pituitary extract and 10^{-9} M cholera toxin. Cultures are maintained at 37 °C in an atmosphere of 5% CO_2/95% air.

Under these conditions, cells explant out from the germinative epithelium to form not typical epithelial cobblestone colonies, but colonies of small rounded cells. This suggests that these are a separate population of epithelial cells, and it is reasonable to assume they are germinative matrix cells in primary culture, but no biochemical markers of differentiation, such as hair-specific keratins, have been identified, nor have any systematic studies of long-term culture or cryopreservation been performed.

2 The sebaceous gland

Isolation of human sebaceous glands

Microdissection

Sebaceous glands can be isolated by careful dissection, but this is difficult as the glands are delicate and difficult to visualise, being of a similar colour to the dermis. For systematic study, therefore, or for large or rapid yields, shearing is the technique of choice (Kealey, 1990). However, there are certain body sites where shearing fails to yield intact, free glands. These unpropitious sites include the scalp and the face, and shearing probably fails because the scalp dermis is too thick and the facial glands too small.

Shearing

Sagittal, frontal, midline chest skin (5 mm×60 mm) obtained from patients undergoing cardiac surgery is a good source of large numbers of human sebaceous glands that may be isolated by shearing. Before obtaining skin, ethics committee permission must be obtained, as must the consent of the patient. These are not usually difficult to obtain as the removal of a sliver of skin from the incision neither affects wound healing nor produces a worse scar.

It is important that the incision is made with a scalpel, as diathermy may damage the glands. The sliver of skin should then be placed in a suitable buffer such as PBS^+ or EBSS. If the media are bicarbonate buffered the containers should be airtight. Bicarbonate-buffered media are preferred because they seem to promote better retention of glandular lipogenesis, possibly due to the bicarbonate dependence of lipogenesis.

In the laboratory, subcutaneous fat is trimmed from the skin and discarded. The skin is then washed four times in sterile EBSS to elute the bactericidal solution that surgeons apply to it, cut into small pieces (<5 mm in length) and sheared in 10 ml of PBS^+, by the repeated action of very sharp scissors, until a porridge-like consistency is obtained (bicarbonate-buffered medium cannot be used as it will be exposed to the air for some time). Samples of this suspension are then diluted in a petri dish with PBS^+ to facilitate microscopy, and the skin glands identified under a binocular microscope. Glands are isolated using watchmaker's forceps (no. 5) taking great care not to damage them. Isolated glands are placed in fresh EBSS until required for experiments.

Sometimes glands are not completely isolated by shearing, but remain attatched to lumps of collagen. They can be easily teased away from the collagen with watchmakers forceps, taking care to pull at the collagen rather than grasp the glands. It is important to use medium at room temperature; ice-cooled medium seems to impair glandular viability, possibly for the same reason that adipocytes are killed at 4 °C, namely that the intracellular fat or sebum solidifies and so ruptures the cell.

Organ maintenance of isolated human sebaceous glands

Glandular lipogenesis *in vitro* is much greater after the overnight maintenance of the glands at 37 °C in bicarbonate-buffered medium than in freshly isolated glands, and this is most probably due either to the bicarbonate dependence of lipogenesis or the temperature sensitivity of the glands. We therefore, as a routine, maintain our glands overnight at 37 °C on sterile nitrocellulose or Millipore filters, pore size 0.45 μm, floating on 5 ml of phenol red-free Williams E medium supplemented with 2 mM L-glutamine, 10 μg/ml insulin, 10 μg/ml transferrin, 10 ng/ml hydrocortisone, 10 ng/ml sodium selenite, 3 nM triiodothyronine, trace elements mix (GIBCO), 100 U/ml^{-1} penicillin, 100 μg/ml streptomycin and 2.5 μg/ml amphotericin B in an atmosphere of 5% CO_2/95% air. Glands maintained for longer periods are also cultured on nitrocellulose filters in the same medium, which is changed every other day.

Phenol red-free medium sustains higher rates of lipogenesis, and better differentiation, than medium containing phenol red. This may be attributed to the oestrogen-like actions of phenol red, since oestrogens inhibit lipogenesis in sebaceous glands. Apart from its role as an indicator, phenol red also protects medium from damage by light, so phenol red-free Williams E medium must be handled in the dark, which is achieved by wrapping the bottles in two layers of aluminium foil. Although it is impossible to perform tissue culture in complete darkness, it is important to expose the medium to as little light as possible by turning out the lights in the tissue culture room and flow hood. The petri dishes are also covered with foil. Nonetheless, a bottle of phenol red-free medium will only keep for 7 days and must be replaced after this time.

The primary culture of sebocytes

There are two different approaches to the primary culture of sebocytes: (i) by isolating sebaceous glands, or (ii) by using a keratotome.

Primary culture of sebocytes from isolated sebaceous glands

Primary explants can be generated from glands isolated by shearing (Kealey, 1990) or by microdissection (Laurent *et al.*, 1992) or by dispase digestion (Xia *et al.*, 1989).

Primary culture of sebocytes from sebaceous glands isolated by shearing or microdissection

Glands can be isolated by microdissection or shearing as described above, and then plated out as described below for glands isolated by dispase digestion, and a high percentage of glands will attach and generate primary explants (our unpublished observations). Attachment can be assisted by first maintaining the glands in a thin layer of medium for 24 h, and both attachment and explant can be assisted by first exposing the glands for a short period to a digestive enzyme such as 20 min in 1 mg/ml collagenase in PBS at 37 °C (Laurent *et al.*, 1992). Although the published methods employ 3T3 feeder layers (see below) primary explants can be established directly onto plastic (our unpublished observations).

Primary culture of sebocytes from sebaceous glands isolated by dispase

If biopsies of whole skin are exposed for 20 h to 2.4 U dispase in PBS^- at 4 °C, or under similar conditions to other digestive enzymes such as thermolysin or trypsin, the dermal/epidermal junction is so weakened that the epidermis can then be peeled off with forceps. The skin glands and appendages then come away with the epidermis, which makes their subsequent microdissection extremely easy (Xia *et al.*, 1989). As the digestion has also weakened the attachment of the glandular and appendigeal basal epithelial cells to each other, and to their basement membranes, glands and appendages thus isolated will subsequently attach and explant easily. (But the cellular architecture will be so disrupted that subsequent organ maintenance will be impossible as the glands or appendages will disintegrate; this applies even to glands, appendages or epidermis isolated by thermolysin digestion which, in our hands, does not spare the basal epithelial cells of the epidermis or glands or appendages (our unpublished observations).)

Primary explants of sebocytes can be established if five glands are plated out on 3T3 feeder layers (2×10^4 cells/cm^2) in 25 cm^2 flasks and covered with culture medium consisting of DMEM containing 4.5 g/l glucose and Ham's F-12 medium (3:1) supplemented with 10% FCS, 100 U/ml penicillin, 100 μg/ml EGF, 0.4 μg/ml hydrocortisone, 10^{-9} M cholera toxin and 3.4 mM L-glutamine. Cultures are maintained at 37 °C in an atmosphere of 95% air/5% CO_2 and the medium is changed every 3 days. The feeder layer is renewed every other week by selectively removing the 3T3 cells with 0.02% EDTA.

These sebocytes may be subcultured. They are detached with 0.25% trypsin and 0.02% EDTA in PBS at 37 °C for 15 min. Trypsin is inhibited by the addition of culture medium. Sebocytes are replated at a density of 4×10^3 cells/cm^2 on a 3T3 feeder layer. By the third passage, however, rates of sebocyte proliferation fall to about 25% of the rate seen at the first passage. No studies on cryopreservation have been performed.

Primary culture of sebocytes isolated by keratatome

An alternative method of isolating and culturing sebocytes was developed by Doran *et al.* (1991). It involves the isolation of a layer of dermis from facelift skin with a keratatome. The top 0.4 mm of skin, containing the epidermis and the some of the upper dermis, is cut with a keratatome and discarded. The next 0.4 mm, containing most of the dermis, is cut and retained. Facial dermis is so rich in sebaceous glands that the predominant cell type in this layer is sebocytic (though facial sebaceous glands themselves are, individually, so small as to preclude shearing for their isolation; see above).

The dermal layer is then incubated in 10 mg/ml dispase (to digest the sebocytes) in DMEM containing 100 U/ml penicillin, 100 μg/ml streptomycin and 10% FCS for 30 min at 37 °C. The sections are then incubated in 0.3% trypsin, 1% EDTA in PBS^- for 15 min at 37 °C, and washed three times in PBS^-. The tissue is then placed in Iscove's medium containing 2% human serum, 8% FCS, 2 mM L-glutamine, 100 U/ml penicillin, 100 μg/ml streptomycin and 4 μg/ml dexamethasone, and scraped with a scalpel blade. This releases the sebocytes, which are then placed at a density of 2×10^4 cells/cm^2 onto 3T3 feeder layers, and cultured in Iscove's medium supplemented as described above.

Surprisingly, this technique appears to yield unmixed cultures of sebocytes, as judged by specific cell markers of differentiation (see below), even though the dermis contains a mixture of cells. It must be assumed that fibroblast growth is inhibited by the 3T3 feeder layer, and that the growth of ORS keratinocytes, which would be the major non-sebocyte epithelial cells, is inhibited by the physiological concentration of calcium and by the lack of EGF.

The keratotome technique yields a large number of cells, which is why Doran *et al.* (1991) developed it, as they were screening large numbers of retinoids for drug development, but most research requirements would probably be better met by the primary culture of isolated sebaceous glands, as the cell purity of the initial tissue would be more certain.

Markers of sebocyte differentiation

The three groups of scientists who have developed the three different methods of culturing sebocytes described above have used similar markers to confirm the cell-specificities of their cultured sebocytes. Keratin 4 is, in skin, a sebocyte-specific marker, and both Laurent *et al.* (1992) and Xia and colleagues (Zouboulis *et al.*, 1991a) demontrated its presence in their primary cultures of sebocytes and its absence in control cultures of interfollicular keratinocytes. Zouboulis *et al.* (1991*a*) did, however, report the expression of keratin 19 by primary cultures of sebocytes, which argues for some de-differentiation on culture. Both Doran *et al.* (1991) and Laurent *et al.* (1992) showed that few of their cultured sebocytes developed cornified envelopes on confluence (0.23% and 0.68% respectively) compared with control cultures of interfollicular keratinocytes (8.8% and 3.2% respectively). Doran *et al.* (1991) moreover showed that only 4.5% of their confluent sebocytes could be induced to cornify with exposure to calcium, as opposed to 72% of keratinocytes.

None of these groups has, however, apparently reported the expression of sebaceous gland antigen (SGA) in their cultures, and though Doran *et al.* (1991), Xia *et al.* (1989) and Zouboulis *et al.* (1991*b*) showed that their sebocytes synthesised more lipid of a sebaceous type (namely rich in squalene, free fatty acids, wax and cholesterol esters) compared with keratinocyte controls, the rates and patterns of lipogenesis seen in primary cultures of sebocytes fall markedly below those seen in organ-maintained sebaceous glands *in vitro*.

Isolation of human sebaceous pilosebaceous infundibulae

The sebaceous pilosebaceous infundibulae are the structures involved in acne.

Isolation by keratotome and microdissection

Infundibulae can be isolated from non-hair-bearing facial skin taken from women undergoing facelift surgery (Guy *et al.*, 1993). Using a keratotome, the top 0.1 mm of the skin containing the epidermis is removed. Then, with the keratotome set to 0.2 mm, the upper portion of the dermis above the sebaceous gland is removed. It is in this layer that the sebaceous infundibulae are located. This layer is placed in sterile PBS^+ and examined under the dissecting microscope. On transillumination the sebaceous infundibulae can be easily identified, as they are much larger than the infundibulae of vellus follicles and they lack the prominent hair of the terminal hair follicle. Moreover, the infundibulae also contain large quantities of sebum, which appears dark on transillumination. Infundibulae are isolated by gentle microdissection with a scalpel and placed in sterile PBS. Some 20–30 infundibulae are generally isolated from an individual where infundibulae are present, but only 50% of subjects appear to possess obvious infundibulae in the retro-auricular skin that is removed at facelift. This percentage probably coincides with that half of the population who have more than average rates of sebum secretion and who, as adolescents, were prone to acne.

Organ maintenance

Infundibulae are maintained free floating, in a multiwell, in keratinocyte SFM (GIBCO), supplemented with 50 μg/ml bovine pituitary extract, 100 U/ml penicillin and streptomycin and 2.5 μg/ml amphotericin B at 37 °C in a humidified atmosphere of 5%CO_2/95% air. Culture medium is changed every other day. To prevent infundibulae from adhering to the surface of the multiwell, it is gently agitated every day. Keratinocyte serum-free medium contains Ca^{2+} at a concentration of 0.09 mM, and this is further supplemented with $CaCl_2.10H_2O$ to a final, physiological Ca^{2+} concentration of 2 mM to retain infundibular morphology.

Primary cell culture and markers of differentiation

We have shown (unpublished results) a similar protocol for the primary culture of hair follicle ORS keratinocytes will also generate a primary culture of infundibular keratinocytes. These cultured infundibular cells have yet to be fully characterised. We have yet to study their long-term culture or cryopreservation.

Freshly isolated, the infundibulum expresses keratins 1, 5, 6, 10, 14, 16 and 17 (as determined by two-dimensional gel electrophoresis; our unpublished results) which compares well with those keratins reported for the upper ORS of the terminal human hair follicle, an analagous structure.

3 Sweat glands

Isolation of human sweat glands

Shearing

Both types of sweat gland can be easily isolated by shearing, as described above for the sebaceous gland (Kealey, 1990). A mid-line chest incision, 15 cm×2.5 mm for a heart operation, could be expected to yield 100 eccrine glands in 30 min; an axillary incision 1.2 cm×2.5 mm (generally obtained at breast cancer staging) will yield some 40 apocrine sweat glands in 20 min.

Microdissection

Both types of gland can also be isolated by microdissection, but this is a laborious procedure which might be justified for physiological studies of transepithelial parameters. Yet glands that have been isolated by shearing can also be microperfused, and no formal comparison between the two techniques, of shearing and microdissection, has yet been performed to determine which yields the optimal electrophysiological preparation.

For cell culture and biochemical experiments, shearing is the isolation method of choice.

Enzymatic separation of dermis and epidermis

The enzymatic separation of dermis and epidermis (described above) will also yield eccrine sweat gland ducts (but not coils, which are retained by the dermis) which will explant easily (Uchida *et al.*, 1993). Such ducts, however, cannot readily be used for physiological studies or for organ maintenance, or for any experiment requiring undamaged cells.

Sweat gland organ maintenance

Both eccrine and apocrine sweat glands can be successfully maintained as whole organs with no apparent loss of function for at least 10 days. This is achieved by maintenance on porous polycarbonate filters floating at the air–surface interface of supplemented Williams E medium at 37 °C in 5% CO_2 /95% air. The only improvement we would recommend over the previously published methods is that FCS should be omitted from the medium, as it promotes fibroblast proliferation but does not improve epithelial cell viability or function.

Primary cell culture

Whole eccrine sweat glands can be cultured, but it is more usual first to microdissect their secretory coils or collecting ducts to obtain cell-type-specific cultures. As the shearing of apocrine sweat glands yields coils that are wholly secretory in nature (see below), their prior microdissection is not necessary.

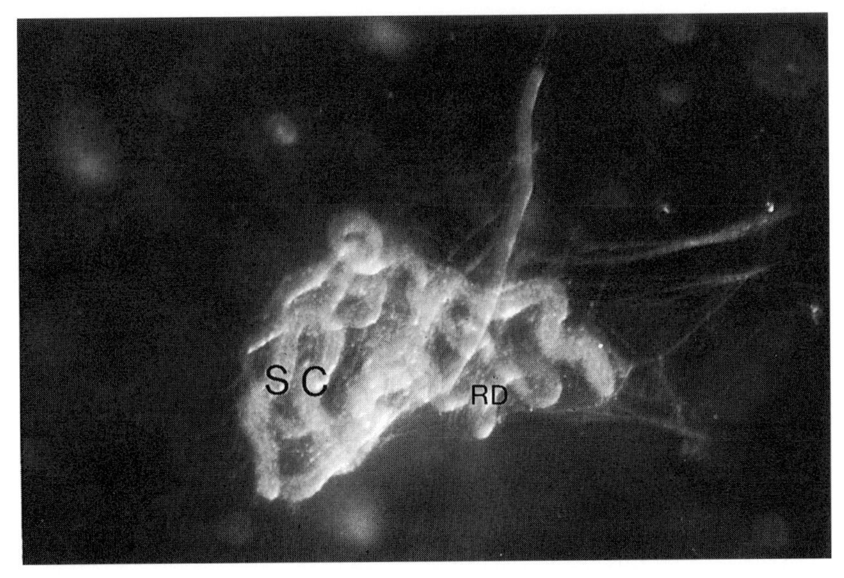

Fig. 8.2. A human eccrine sweat gland isolated by shearing. The reabsorptive duct (RD) and secretory coil (SC) can be distinguished as the coil is larger in diameter, and its borders less clearly defined. ×45. (Reproduced with permission from the *Biochemical Journal*, **212**, 143–8, 1983.)

Microdissection of eccrine sweat glands

Anatomically, eccrine sweat glands consist of a coil and a duct. Physiologically, eccrine sweat glands consist of a secretory coil (which secretes the primary secretion) and a reabsorptive duct (which reabsorbs up to half the NaCl from the primary secretion). All the physiological secretory coil, and much of the reabsorptive duct, are contained within the anatomical coil; the straight duct that emerges from the coil, leading up to the acrosyringium at the skin surface, is thought to be largely a conduit as it is depleted of mitochondria, though rich in tonofilaments.

The secretory coil and the reabsorptive duct can easily be distinguished from each other within the anatomical coil by transillumination: the coil is broader and possesses an indistinct margin, while the reabsorptive duct is narrower, of a waxy appearance and possesses a more marked lumen (Fig. 8.2).

Glands can be easily separated into coil and duct. First, they are incubated for 20 min in isolation medium containing 2 mg/ml collagenase type II. (Different authors have used different isolation media, including PBS, PBS:Earle's, HEPES-buffered saline, and Williams E medium containing between 1% and 5% FCS; no formal comparison between these different isolation media has been performed, but they would appear to be interchangeable.) The glands are then transferred to enzyme-free isolation medium and dissected under sterile conditions at a magnification of ×30, either using mounted A4 insect pins or sandpaper-sharpened forceps. The glands are carefully manipulated until the free end of the duct can be grasped and pulled on lightly. In most cases this will unravel the duct until the join with the coil is visible. The join can be removed to avoid contamination. In some cases the duct is entangled more firmly and the whole gland must be dissected out to ensure complete separation.

For physiological or other experiments where prior digestion of the gland is undesirable, glands are dissected by manipulating the attached collagen fibres so as not to touch or damage the epithelial cells.

Apocrine glands isolated by shearing can be microdissected. Shearing yields only secretory coil. We have confirmed through the repeated sectioning of sheared glands that all tubular profiles are coil (i.e. surrounded by myoepithelial cells). No description of the isolation of apocrine duct has yet been made.

Primary sweat gland cell culture

Primary sweat gland cell culture can be made either from explants of whole glands or coils or ducts, or from dissociated cells (Lee *et al.*, 1986; Uchida *et al.*, 1993) (see Fig. 8.3).

Primary sweat gland explant cell culture. Either microdissected eccrine duct and coil or whole glands (treated with collagenase type II for 20 min) are transferred to 25 cm^2 tissue culture flasks (filter-capped flasks from Corning, cat. no. 25103–25, are recommended, as the caps seem to inhibit fungal infection, and the plastic seems to foster good sweat gland cell culture) containing only 1 ml culture medium to aid attachment. After 24 h a further 2–3 ml of medium is carefully added and thereafter the cells are fed every 3 days. A number of different media have been used by researchers, but Mammary Epithelial Growth Medium (Clonetics) appears to give optimal results. It is an MCDB 170 derivative supplemented with 10 ng/ml EGF, 10 μg/ml insulin, 0.5 μg/ml hydrocortisone, bovine pituitary extract (diluted according to the manufacturer's instructions), 50 μg/ml gentamicin and 50 ng/ml amphotericin B. Significant outgrowth occurs from around 80% of eccrine whole glands or coils, and from around 60% of ducts. Cells tend to congregate and primary cultures do not usually reach confluence unless sweat glands are seeded at high density. However, on passage (up to four passages can be achieved) confluence can be reached.

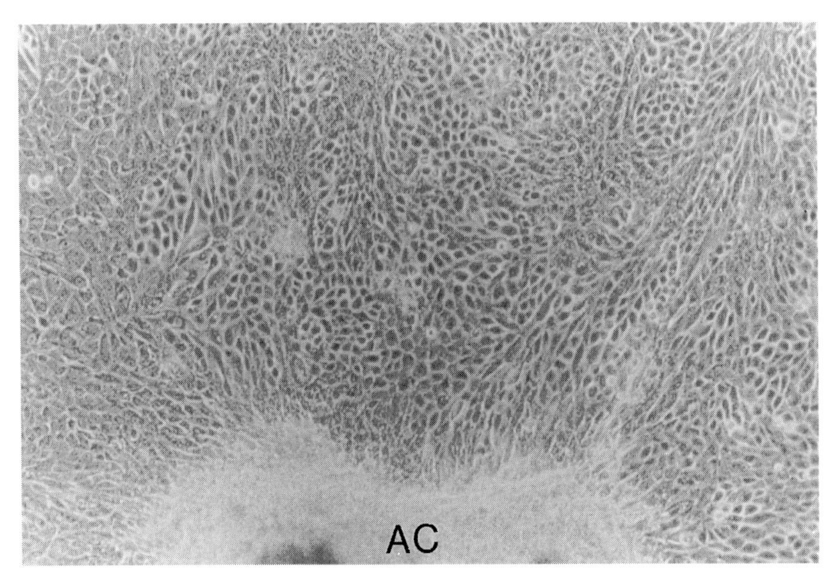

Fig. 8.3. An 11-day-old primary explant culture of human apocrine sweat gland coil cells. The structure of the apocrine coil (AC) has largely disintegrated. The majority of the primary cells demonstrate an epithelial morphology. ×55.

Apocrine coils can be cultured in the same manner (though 40 min of collagenase digestion is optimal), but only 30–50% of coils produce outgrowths.

For electrophysiological studies, cells are best harvested at around 15 days. They are incubated in 2% EDTA for 10 min and then trypsin/EDTA for 5–10 min. Addition of serum-containing medium halts enzyme action. After being spun down for 2 min at 100 *g*, cells are pipetted on to permeable supports (Fig. 8.4) to allow Ussing chamber experiments, or on to coverslips for Fura-2 Ca^{2+} investigations.

Primary sweat gland dispersed cell culture. Microdissected duct and coil cells can be dispersed by multiple collagenase/DNase I digestions, followed by repeated passage through a tungsten electron microscope grid. Duct cells have also been dispersed by overnight incubation in dispase followed by trypsin/EDTA exposure and gentle agitation (Uchida *et al.*, 1993). But these methods yield few cells and the former method is immensely difficult. Dispersed cells, therefore, may be useful for certain discrete physiological studies, but primary explant is recommended for primary culture.

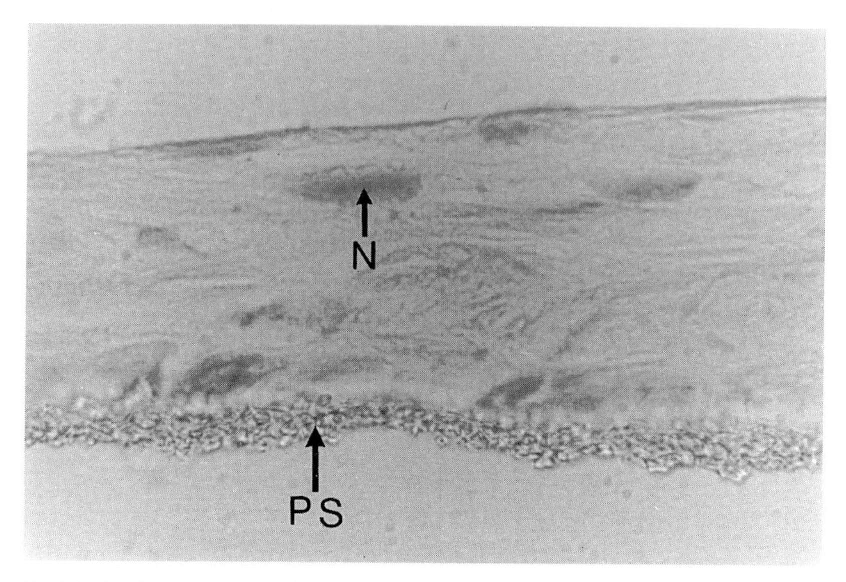

Fig. 8.4. A light micrograph of cultured human apocrine sweat gland coil cells on a Costar Transwell permeable support (PS), 4 days after seeding, stained with phosphotungstenic acid haematoxylin. Flattened nuclei (N) are clearly visible. ×350.

Cytochemical markers of differentiation

Attempts have been made to characterise the keratin intermediate filaments of eccrine sweat glands, by immunostaining and by two-dimensional gel electrophoresis. From staining of fixed sections it appears that cytokeratins 7, 18 and 19 are expressed by the cells of the secretory coil, 19 and 10 or 11 by the luminal cells of the duct, and 5 by the basal cells of the duct. Two-dimensional gel electrophoresis using whole glands suggests, in addition, the presence of cytokeratins 6, 8, 14 and 15. Cultured eccrine coil and duct cells stain positively with an antibody which detects cytokeratins 8, 18 and 19 (Lee *et al.*, 1986), but an exhaustive cytokeratin analysis of eccrine sweat glands *in vivo* and *in vitro* remains to be performed.

Preliminary work on apocrine gland cytokeratins suggests differences between eccrine and apocrine expression. Apocrine glands appear to lack cytokeratins 6 and 15, yet to express 4 in addition to the eccrine complement, but a comprehensive study has yet to be performed.

Electrophysiological markers of differentiation

The major ion movements that underlie fluid secretion by the secretory coil, and reabsorption by the collecting duct, *in vivo*, have been characterised. In brief, secretion is mediated by an electrogenic movement of chloride ions from the basolateral to the luminal poles. This is followed by an electrochemically generated movement of sodium, and an osmotically generated movement of fluid.

Sodium reabsorption is dominated, electrophysiologically, by an electrogenic movement of sodium that is inhibitable by luminally applied amiloride.

On culture, duct cells retain an absorptive phenotype (Pedersen, 1993), which is encouraging; disappointingly, coil cells on culture adopt a similar absorptive phenotype, dominated by an amiloride-inhibitable sodium flux (Brayden *et al.*, 1988). This implies that secretory cells lack a crucial differentiation signal *in vitro*, and so 'default' to an absorptive phenotype.

In cultured apocrine cells, electrophysiological de-differentiation also occurs, though this has not been fully characterised.

Immortalised sweat gland cell lines

Primary-cultured human eccrine sweat gland cells have been immortalised using the SV40 virus, or an SV40/Ad5 chimera (Lee & Dessai, 1989; Buchanan *et al.*, 1990; Bell & Quinton, 1995). In general, this results in aneuploid cells with some phenotypic changes and life-spans ranging from 20 to 250 population doublings. A cell line has been created from primary cultured eccrine duct cells (Bell & Quinton, 1995) which maintains an amiloride-sensitive sodium conductance and a physiologically appropriate chloride conductance, as well as a typically epithelioid cobblestone appearance. Attempts to immortalise cultured eccrine secretory coil cells have been less successful. In one case, cholinergic sensitivity was lost (Lee & Dessai, 1989), whilst another developed domes in culture (Buchanan *et al.*, 1990), indicating a loss of secretory activity and the development of an absorptive phenotype.

Horse sweat gland cells, unlike those from the human, can spontaneously immortalise on culture (Wilson *et al.*, 1993). They have yet to be fully characterised by their cytokeratins or electrophysiology.

Cryopreservation

All immortalised sweat gland cell lines can be preserved under liquid nitrogen using standard techniques (see above). To our knowledge, no non-immortalised sweat gland cell lines have been so preserved.

Conclusion

In summary, the major human skin glands and appendages can now be isolated. Their organ maintenance is gratifyingly successful, but their cell culture is still in its infancy. Most relevant cell types can be cultured, and some have been immortalised, but de-differentiation on culture is normal. The ultimate ambition of successful histiotypic culture, and of generating immortal cell lines that retain fully differentiated phenotypes, is still far from being achieved.

Acknowledgements

We thank Unilever Research for financial support.

References

Bell, C.L. & Quinton, P.M. (1995). An immortal cell line to study the role of endogenous CFTR in electrolyte absorption. *In Vitro Cell Dev. Biol.*, **31A**, 30–6.

Brayden, D.J., Cuthbert, A.W. & Lee, C.M. (1988). Human eccrine sweat gland epithelial cultures express ductal characteristics. *J. Physiol.* (*Lond.*), **405**, 657–75.

Buchanan, J.A., Yeger, H., Tabcharani, J.A., Jensen, T.J., Auerbach, W., Hanrahan, J.W., Riodan, J.R. & Buchwald, M. (1990). Transformed sweat gland and nasal epithelial cell lines from control and cystic fibrosis individuals. *J. Cell Sci.*, **95**, 109–123.

Cotsarelis, G., Sun, T.-T. & Lavker, R.M. (1990). Label-retaining cells reside in the bulge area of pilosebaceous units: implications for follicular stem cells, hair cycle and skin carcinogenesis. *Cell*, **61**, 1329–37

Doran, T.I., Baff, R., Jacobs, P. & Pacia, E. (1991) Characterisation of human sebaceous sebaceous cells *in vitro*. *J. Invest. Dermatol.*, **96**, 341–8.

Guy, R., Ridden, C., Barth, J.H. & Kealey, T. (1993). Isolation and maintenance of the human pilosebaceous duct: 13-*cis*-retinoic acid acts directly on the duct *in vitro*. *Br. J. Dermatol.*, **128**, 242–8.

Hennings, H., Michael, D., Cheng, C., Steinert, P., Holbrook, K. & Yuspa, S.H. (1980). Calcium regulation of growth and differentiation of mouse epidermal cells in culture. *Cell*, **19**, 245–54.

Kanazaki, T., Kanamaru, T., Nishiyama, S., Eto, H., Kobayash, H. & Hashimoto, K. (1983). Three dimensional hair follicular differentiation of a trichilemmoma cell line *in vitro*. *Dev. Biol.*, **99**, 324–30.

Kealey, T. (1990) Effects of retinoids on human sebaceous glands isolated by shearing. *Methods Enzymol.*, **190**, 338–45.

Laurent, S.J., Mednicks, M.I. & Rosenfield, R.L. (1992). Growth of sebaceous cells in monolayer culture. *In Vitro Cell. Dev. Biol.*, **28A**, 83–9.

Lee, C.M. & Dessai, J. (1989). NCL-SG3: a human eccrine sweat gland cell line that retains the capacity for transepithelial ion transport. *J. Cell Sci.*, **92**, 241–9.

Lee, C.M., Carpenter, F., Coaker, T. & Kealey, T. (1986). The primary culture of epithelia from the secretory coil and collecting duct of normal human and cystic fibrotic eccrine sweat gland. *J. Cell Sci.*, **83**, 103–18.

Limat, A., Breitkreutz, D., Hunziker, T., *et al.* (1991). Restoration of the epidermal phenotype by follicular outer root sheath cells in recombinant culture with dermal fibroblasts. *Exp. Cell Res.*, **194**, 218–27.

Limat, A., Wyss-Coray, T., Hunziker, T. & Braathen, L.R. (1994). Comparative analysis of surface antigens in cultured human outer root sheath cells and epidermal keratinocytes: persistence of low expression of class 1 MHC antigens in outer root sheath cells *in vitro*. *Br. J. Dermatol.*, **131**, 184–90.

Messenger, A.G. (1984). The culture of dermal papilla cells from human hair follicles. *Br. J. Dermatol.*, **110**, 685–9.

Pedersen, P.S. (1993). Human reabsorptive sweat duct in primary cell culture. *Dan. Med. Bull.*, **40**, 208–23.

Philpott, M.P., Green M.R. & Kealey, T. (1990) Human hair growth *in vitro*. *J. Cell Sci.*, **97**, 463–71.

Reynolds, A.J., Lawrence, C.M. & Jahoda, C.A.B. (1993). Human hair germinative epidermal cell culture. *J. Invest. Dermatol.*, **101**, 634–8.

Stark, H.-J., Breitkreutz, D., Limat, A., Bowden, P. & Fusenig, N.E. (1987). Keratins of the human hair follicle; hyperproliferative keratins are consistently expressed in outer root sheath cells *in vivo* and *in vitro*. *Differentiation*, **35**, 236–48.

Uchida, N., Oura, H., Nakanishi, H., Urano, Y. & Arase, S. (1993). Dispersed cell culture of human sweat duct cells under serum-free conditions. *J. Dermatol.*, **20**, 684–90.

Westgate, G.E., Gibson, W.T., Kealey, T. & Philpott, M.P. (1993). Prolonged growth of human hair follicles *in vitro* in a serum free defined medium. *Br. J. Dermatol.* **129**, 372–9

Wilson, S.M., Pediani, J.D., Ko, K.-W., Bovell, D.L., Kitson, S., Montgommery, I., Brown, U.M.O., Smith, G.L., Elder, H.Y. & Jenkinson, D.McE. (1993). Investigation of stimulus–secretion coupling in eccrine sweat gland epithelia using cell culture techniques. *J. Exp. Biol.*, **183**, 279–99.

Xia, L., Zouboulis, C., Detmar, M., Mayer-da-Silva, A., Stadley, R. & Orfanos, C.E. (1989). Isolation of human sebaceous glands and cultivation of sebaceous gland-derived cells as an *in vivo* model. *J. Invest. Dermatol.*, **93**, 315–21.

Zouboulis, C.C., Xia, L., Detmar, M., *et al.* (1991*a*). Culture of human sebocytes and markers of sebocytic differentiation *in vitro*. *Skin Pharmacol.*, **4**, 74–83.

Zouboulis, C.C., Korge, B., Akamatsu, H., *et al.* (1991*b*). Effect of 13-*cis*-retinoic acid, all *trans* retinoic acid, and Acitrenin on the proliferation, lipid synthesis and keratin expression of cultured human sebocytes *in vitro*. *J. Invest. Dermatol.*, **96**, 792–7.

Index

Page numbers in *italic* type refer to illustrations.